Atomic Force Microscopy

PETER EATON

Requimte, and Faculty of Science,
University of Porto

PAUL WEST

AFMWorkshop, Inc.

OXFORD

UNIVERSITY PRESS

OXFORD
UNIVERSITY PRESS

Great Clarendon Street, Oxford OX2 6DP

Oxford University Press is a department of the University of Oxford.
It furthers the University's objective of excellence in research, scholarship,
and education by publishing worldwide in

Oxford New York

Auckland Cape Town Dar es Salaam Hong Kong Karachi
Kuala Lumpur Madrid Melbourne Mexico City Nairobi
New Delhi Shanghai Taipei Toronto

With offices in

Argentina Austria Brazil Chile Czech Republic France Greece
Guatemala Hungary Italy Japan Poland Portugal Singapore
South Korea Switzerland Thailand Turkey Ukraine Vietnam

Oxford is a registered trade mark of Oxford University Press
in the UK and in certain other countries

Published in the United States
by Oxford University Press Inc., New York

British Library Cataloguing in Publication Data

Data available

Library of Congress Cataloging in Publication Data

Data available

Typeset by SPI Publisher Services, Pondicherry, India
Printed in Great Britain
on acid-free paper by
Clays Ltd, St Ives plc

ISBN 978–0–19–957045–4 (Hbk.)

5 7 9 10 8 6

Peter Eaton: To Maria – thanks for all the help and support while I was doing this

Paul West: To Christoph, Gerd, Heini, Cal – thanks for a wonderful gift

Contents

Preface

The aim of this book is to demystify AFM. When you've read this book, you should understand how AFM works, including the main modes of operation, how to make measurements with an AFM, how to optimize your measurements, how to analyse your data, how to spot and to avoid problems with it, and you should have a good idea of what AFM is useful for.

This book was written so that the reader can dip in and out of the book, and that the chapters will be – more or less – readable independently, but the book will make most sense read from start to finish. Certainly if you know nothing about AFM yet, you will get the most out of this book if you read it all the way through, in the right order. But if you already know how the technique works and just want to analyse data, go straight to the chapter on image analysis; it will be perfectly readable without reading the prior sections. We assume no prior knowledge about AFM. This book is designed to be readable to someone with a freshman college-level of education, and an interest in AFM. On the other hand, some of the sections are highly detailed, and we expect that even experienced AFM operators will find a lot of useful information in them.

The first chapter introduces AFM, and places it in the context of the preceding techniques, as well as how it compares with other microscopy techniques. The second chapter describes how modern AFMs are built, and how they work. Even if you are an experienced AFM user there may well be details in Chapter 2 that you are not aware of. Knowledge of how the instrument works can greatly improve your use of it, and we hope that without going into too great technical details, this chapter has all the information an AFM user could need about how AFMs work, and importantly, why they work that way.

The third chapter then describes the major AFM modes in use. We discuss the way the modes work, and what information they can give, as well as the advantages and disad-vantages of the different modes available. After describing the modes used to collect sample topography (i.e. imaging modes), modes used to get other information about the sample are described, for example how to use AFM to get thermal, magnetic, and mechanical information about a sample's surface. AFM can also be used to record information about how individual molecules interact, and even how protein unfolding can be measured. All these modes are extensively referenced, and there are examples of each in the last chapter as well, showing a typical application of these modes in use.

In Chapter 4 we describe how to measure AFM images. If you have already measured images, you might be tempted to skip this chapter, but it may still be worth a look, because almost every user of AFM measures their images in a slightly different way, and you may well find some useful tips here. Particularly, we show examples of how you can use the information in the preceding chapters to understand why your images are good (or not so good). We show how to optimize scanning conditions, for the best resolution, image quality, and accuracy. This information should not be seen as the replacement for your instrument manual, but a complement to it. In combination with the other information in this book, this chapter should help you to understand more deeply the process of scanning AFM images, so you can get better, more reproducible images.

Even the best data needs the right treatment to get the most useful information out of it, and Chapter 5 is all about how to process, display, and analyse AFM data. This chapter will be particularly useful if you have AFM data provided to you by an instrument operator which you did not collect yourself. Initially AFM analysis software can be very confusing, as there are so many different operations you can carry out, some of which may permanently alter your data. It is important to only apply those operations which are useful for your application, and avoid 'over-processing' of your data. This chapter will show how to maintain data integrity, and how to optimize and process the data for best effect.

Chapter 6 shows how to spot common artefacts in AFM images. Like all scientific measurements, AFM is prone to a number of artefacts, and unless you already know your sample very well, they can be quite tricky to spot. After years of usage this becomes second nature, and certain recurring artefacts will be obvious when they occur. But some rare artefacts can be easily missed, and new AFM users have little chance of knowing when an image has something wrong with it. The artefacts can come from the tip, from the environment, or be inherent in the technique itself. In this chapter, we give examples of the common image artefacts, and describe what you can do to avoid them.

It is obvious that AFM is not the solution to *all* scientific and technical problems; it does have some disadvantages, and sometimes other microscopy techniques are more appropriate for a particular problem. However, AFM has been applied with great success to an incredibly wide range of scientific and technological fields, and in the final chapter we present a range of applications that illustrate the breadth and depth of the uses of AFM.

Chapter 1

Introduction

Atomic force microscopy is an amazing technique that allows us to see and measure surface structure with unprecedented resolution and accuracy. An atomic force microscope (AFM) allows us, for example, to get images showing the arrangement of individual atoms in a sample, or to see the structure of individual molecules. By scanning in ultra-high vacuum at cryogenic temperatures the hopping of individual atoms from a surface has been measured [1]. On the other hand, AFM does not need to be carried out under these extreme conditions, but can be carried out in physiological buffers at 37 °C to monitor biological reactions and even see them occur in real time [2–4]. Very small images only 5 nm in size, showing only 40–50 individual atoms, can be collected to measure the crystallographic structure of materials, or images of 100 micrometres or larger can be measured, showing the shapes of dozens of living cells at the same time [5–9]. AFM has a great advantage in that almost any sample can be imaged, be it very hard, such as the surface of a ceramic material, or a dispersion of metallic nanoparticles, or very soft, such as highly flexible polymers, human cells, or individual molecules of DNA. Furthermore, as well as its use as a microscope, which is to say as an imaging tool, AFM has various 'spectroscopic' modes, that measure other properties of the sample at the nanometre scale. Because of this, since its invention in the 1980s, AFM has come to be used in all fields of science, such as chemistry, biology, physics, materials science, nanotechnology, astronomy, medicine, and more. Government, academic and industrial labs all rely on AFM to deliver quantitative high-resolution images, with great flexibility in the samples that can be studied.

An AFM is rather different from other microscopes, because it does not form an image by focusing light or electrons onto a surface, like an optical or electron microscope. An AFM physically 'feels' the sample's surface with a sharp probe, building up a map of the height of the sample's surface. This is very different from an imaging microscope, which measures a two-dimensional projection of a sample's surface. Such a two-dimensional image does not have any height information in it, so with a traditional microscope, we must infer such information from the image or rotate the sample to see feature heights. The data from an AFM must be treated to form an image of the sort we expect to see from a microscope. This sounds like a disadvantage, but the treatment is rather simple, and furthermore it's very flexible, as having collected AFM height data we can generate images which look at the sample from any conceivable angle with simple analysis software. Moreover, the height data makes it very simple to quickly measure the height, length, width or volume of any feature in the image.

The fact that the AFM operates differently from most microscopes, and that the AFM probe physically interacts with the sample, means however that it is not as intuitive to use as optical microscopes. While most people understand the basic principles of light microscope use, i.e. focusing, illumination, depth of field, and so on, the use of AFM

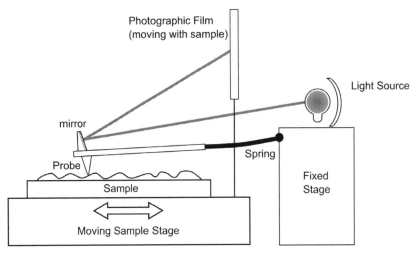

Fig. 1.1. Optical lever design used for one of the early models of a surface profiler in the 1920s. This profiler had a vertical resolution of approximately 25 nm.

has none of these concepts. There is nothing to focus, there's no illumination of the sample, and zero depth of field, so operation of an AFM is rather different from many users' expectations of a microscope. This means that both operation of and understanding the data from an AFM can be initially confusing. However, the principles, which will be explained in the following chapters, are really rather simple and having grasped these, both data analysis and acquisition will become much more intuitive. Like all scientific techniques, atomic force microscopy was a development of previously known methods, but is a technique which led to a revolution in microscopy. The development of AFM from these earlier techniques is discussed in the next section.

1.1   Background to AFM

As mentioned above, the AFM works by scanning a probe over the sample surface, building up a map of the height or topography of the surface as it goes along. It was not the first instrument to work in this way however. The predecessor of the AFM was the stylus profiler, which used a sharp tip on the end of a small bar, to which was dragged along the sample surface, and built up a map, or more often a linear plot, of sample height. An example of an early profiler is shown in Figure 1.1. This profiler, described by Shmalz in 1929, utilized an optical lever to monitor the motion of a sharp probe mounted at the end of a cantilever [10]. A magnified profile of the surface was generated by recording the motion of the stylus on photographic paper. This type of 'microscope' generated profile 'images' with a magnification of greater than $1000 \times$.

A common problem with stylus profilers was the possible bending of the probe from collisions with surface features. Such 'probe bending' was a result of horizontal forces on the probe caused when the probe encountered large features on the surface. This problem was first addressed by Becker [11] in 1950. Becker suggested oscillating the probe from an

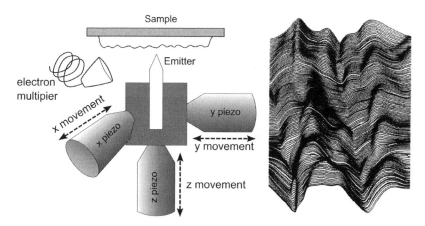

Fig. 1.2. A schematic diagram of Young's topografiner (left), and one of the first images collected with the instrument (right). Reprinted with permission from [12].

initial position above the surface to approach contact with the surface. Becker remarked that when using this vibrating profile method for measuring images, the detail of the images would depend on the sharpness of the probe. Stylus profilers are still in use today, and have developed considerably. However, fundamental problems with this sort of instrument persist, notably that the probe touches the surface in an uncontrolled way, which can lead to probe damage in the case of a hard sample, and sample damage in the case of a soft sample. Either of these problems would reduce the fidelity of the image obtained, as well as the resolution achievable.

In 1971 Russell Young demonstrated a non-contact type of stylus profiler [12]. In his profiler, called the topografiner, Young used the fact that the electron field emission current between a sharp metal probe and a surface is very dependent on the probe sample distance for electrically conductive samples. In the topografiner (shown in Figure 1.2), the probe was mounted directly on a piezoelectric ceramic element which was used to move the probe in a vertical direction (z) above the surface. Further piezoelectric elements moved the probe in the other axes over the sample.

An electronic feedback circuit monitoring the electron emission was then used to drive the z-axis piezoelectric element and thus keep the probe–sample distance at a fixed value. Then, with the x and y piezoelectric ceramics, the probe was used to scan the surface in the horizontal (X-Y) dimensions. By monitoring the X-Y and Z position of the probe, a 3-D image of the surface was constructed. The resolution of Young's topografiner was limited by the instrument's vibrations.

In 1981 Binnig and Rohrer, working at IBM, were able to improve the vibration isolation of an instrument similar to the topografiner such that they were able to monitor electron tunnelling instead of field emission between the tip and the sample. This instrument was the first scanning tunnelling microscope (STM) [13–15]. A schematic diagram of the STM is shown in Figure 1.3. The STM works by monitoring the tunnelling current and using the signal, via a feedback loop, to keep the STM tip (a sharp metal wire) very close to the sample surface while it is scanned over the surface in the X and Y axes in a

x, y & z control voltages

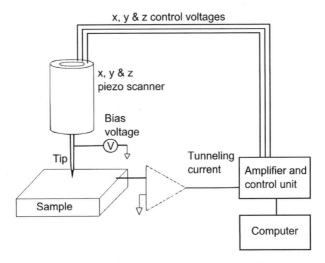

Fig. 1.3. Simplified schematic of a scanning tunnelling microscope (STM).

raster pattern. Like the topografiner, the movement of the tip over the surface in x, y and z is controlled with three piezoelectric elements (in Figure 1.3, the three elements are integrated together in a tube structure; this is discussed further in Chapter 2). The distance the z piezo has to move up and down to maintain the tunneling current at the same value is equivalent to the sample height, so the computer can build up a map of sample height as the tip scans over the surface. The reason the instrument was so much more successful than the topografiner is that electron tunnelling is much more sensitive to tip–sample distance than field emissions, so the probe could be scanned very close to the surface. In fact, the probability of electron tunnelling is so strongly dependent on distance that effectively only the very last atom of the STM tip can undergo tunnelling. Because it is this last atom which is most sensitive to tunnelling from the surface, the structure of the tip far from the surface is not very important, so atomically sharp tips are easy to produce. For their very first experiments, Binnig and Rohrer levitated the entire instrument magnetically to counter vibrations; however later designs did not require this. The results of these early experiments were astounding; Binnig and Rohrer were able to see individual silicon atoms on a surface, [14, 16]. Without the STM, attaining this kind of resolution required a transmission electron microscope (TEM), which weighs thousands of kilograms, and fills a room. Furthermore, when the STM was invented, atomic structure could only be observed indirectly by diffraction patterns, while the STM could do it directly by imaging individual atoms. That the STM could do this when it was only a small instrument, suspended with springs to counter vibrations, seemed incredible, and Binnig and Rohrer later shared the Nobel Prize for physics in 1986 for the invention of the STM [17].

Although the STM was considered a fundamental advancement for scientific research, it had limited applications, because it worked only on electrically conductive samples. Despite these limits, STM remains a very useful technique, and is used widely in particular in physics and materials science to characterize the atomic structure of metals and semiconductors, and for fundamental studies of electronic effects at metal surfaces. Figure 1.4 shows an STM image, illustrating the atomic resolution routinely obtained in STM.

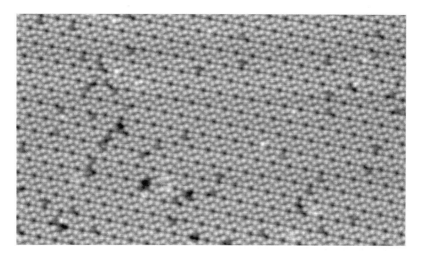

Fig. 1.4. Example of an STM image with atomic resolution. The image shows an atomic-resolution image of the 5 × 5 reconstruction of the Si(111) surface. Individual atoms, defects and vacancies are visible. Reproduced with kind permission from Dr. Randall Feenstra.

Despite the amazing results obtained with STM, the limitation to conducting samples led the inventors to immediately think about a new instrument that would be able to image insulating samples. In 1986 Binnig, Quate and Gerber published a paper entitled 'Atomic Force Microscope' [18, 19]. In that paper they described how they replaced the wire of a tunnelling probe from the STM with a lever made by carefully gluing a tiny diamond onto the end of a spring made of a thin strip of gold. This was the cantilever of the first AFM. Although the first instrument was used only for a few experiments, the results produced had such great impact that the first instrument now resides in the science museum in London. The movement of the cantilever was monitored by measuring the tunnelling current between the gold spring and a wire suspended above it. This set-up was highly sensitive to the movement of the probe as it scanned along the sample, again moved by piezoelectric elements. In their paper, Binnig *et al.* proposed that the AFM could be improved by vibrating the cantilever above the surface [20]. Thankfully nowadays we don't have to glue tiny diamonds onto gold levers to carry out AFM, but this first instrument led to the whole field of AFM. The instrument, and the first image recorded in AFM, are shown in Figure 1.5.

The AFM caused a revolution. Suddenly, with a relatively cheap and simple instrument, extremely high-resolution images of nearly any sample were possible. While initial images, such as that shown in Figure 1.5, did not have as high resolution as STM, atomic-resolution images were soon reported [21]. Soon after the invention of the AFM, the gold leaf/diamond combination was replaced by much more reproducible cantilever manufacture by silicon lithography, which enables the production of more than 400 cantilevers on a single 7-inch wafer [22]. Furthermore, it was quickly realized that simpler methods than the STM could be used to detect the motion of the cantilever. Nowadays, most AFMs use a light lever to sensitively detect the motion of the cantilever, this method is considerably simpler than the STM set-up, allows for larger cantilever motions, and is

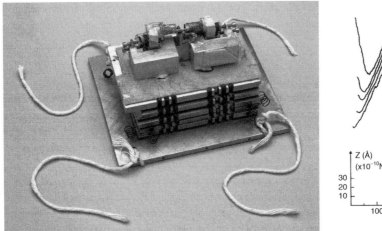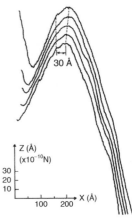

Fig. 1.5. The first AFM instrument built by Binnig, Quate and Gerber in the Science Museum, London (image copyright Science Museum/SSPL), and the first AFM image – reprinted with permission from[19]. Copyright 1986 by the American Physical Society.

still sensitive to sub-angstrom motions of the cantilever [23, 24]. Furthermore, as suggested by Binnig *et al.*, oscillating modes have further increased the range of samples that AFMs can scan, and reduced the chance of sample damage as well.

Due to the high interest in AFM, commercial instruments were soon being produced, the first available from 1988. Together, AFM and STM are often referred to as scanning probe microscopy, or SPM. A further explanation of terminology in the SPM field is given in Chapter 3. Since AFM and STM instruments share several components in common, it is relatively simple to build an instrument capable of carrying out both kinds of microscopy. Since together they are referred to as SPM, and because some instruments perform both STM and AFM, the techniques are often seen as being very similar. However, since its development, AFM has been modified to measure a huge number of different properties, and perform lots of additional (non-imaging experiments), and combined with the techniques' greater flexibility in terms of types of samples scanned, means AFM is today much more widely used than STM. This book concentrates on AFM, and will not discuss STM further. For the reader interested in further details of STM, the works [25, 26] are recommended.

1.2 AFM today

The AFM can be compared to traditional microscopes such as the optical or scanning electron microscopes for measuring dimensions in the horizontal axis. However, it can also be compared to mechanical profilers for making measurements in the vertical axis to a surface. One of the great advantages of the AFM is the ability to magnify in the X, Y and Z axes. Figure 1.6 shows a comparison between several types of microscopes and profilometers. As shown in Figure 1.6, one of the limiting characteristics of the AFM is that it is

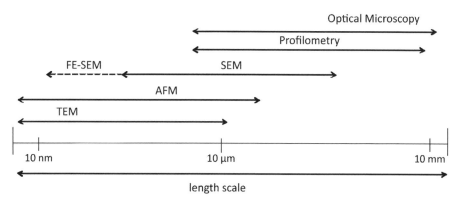

Fig. 1.6. Comparison of length-scales of various microscopes.

not practical to make measurements on areas greater than about 100 μm. This is because the AFM requires mechanically scanning the probe over a surface, and scanning such large areas would generally mean scanning very slowly. Exceptions to this include parallel AFM that measure small areas but with many probes to build up a large dataset, or 'fast-scanning' AFMs, which are discussed in Chapter 2.

When compared to a profiler, the AFM has a greater X-Y resolution because in the AFM the probe is sharper. The fine control of probe–surface forces enabled by this feedback mechanism enables the use of lower loading forces, which allows the use of much sharper probes, resulting in much higher X-Y resolution. The difference in applied force is very high, while profilometers will typically apply *ca.* 10^{-6} N to the surface, AFMs can image with 10^{-9} N or less. Profilers can have high vertical resolutions, as low as 0.5 Å. However, much greater bandwidth in the AFM experiments means that practically, the AFM height resolution is far greater than that of the profilometers. This is because the bandwidth limits on profilometers mean that to achieve high height resolution scanning must occur very slowly.

The length-scale of an optical microscope overlaps nicely with an AFM. Thus, an AFM is often combined with an optical microscope and with this combination it is possible to have a combined field of view with a dynamic range from mm to nm. In practice, a simplified optical microscope, known as an inspection scope, is usually used for selecting the location for AFM scanning. However, a combination of high-resolution optical microscopes, often with fluorescence microscopy integration, with AFM also has great advantages, especially in biology. This is discussed further in Chapter 2 and in Section 7.3. The combination of AFM with other microscopes or instruments is made simple by the AFM's small size.

The AFM is most often compared with the electron beam techniques such as the Scanning Electron Microscope (SEM) or Transmission Electron Microscope (TEM). As may be seen in Figure 1.6, the dimensional range of these techniques is rather similar, with SEM (usually) having a somewhat lower resolution to AFM, while the ultimate resolution of TEM is quite similar to that of AFM. Table 1.1 contains a list of some of the major factors in comparison of AFM with SEM and TEM.

In general, it is easier to learn to use an AFM than an electron microscope because there is minimal sample preparation required with an AFM, and nearly any sample can be

Table 1.1. Comparison of AFM with SEM and TEM.

	AFM	SEM	TEM
Sample preparation	little or none	from little to a lot	from little to a lot
Resolution	0.1 nm	5 nm	0.1 nm
Relative cost	low	medium	high
Sample environment	any	vacuum(SEM) or gas (environmental SEM)	vacuum
Depth of field	poor	good	poor
Sample type	Conductive or insulating	conductive	conductive
Time for image	2–5 minutes	0.1–1 minute	0.1–1 minute
Maximum field of view	100 μm	1 mm	100 nm
Maximum sample size	unlimited	30 mm	2 mm
Measurements	3 dimensional	2 dimensional	2 dimensional

measured. With an AFM, if the probe is good, a good image is measured. Because TEM and SEM usually operate in a vacuum, and require a conductive sample (so non-conductive samples are usually coated with a metallic layer before imaging), AFM has the advantage of being able to image the sample with no prior treatment, in an ambient atmosphere. This makes scanning quicker, and can also mean fewer artefacts are introduced by the vacuum drying, or the coating procedure. On the other hand, AFM image recording is usually slower than an SEM, so if a large number of features on one sample are required, AFM may be considerably slower than SEM for the same sample.

As we will see in the following chapters, AFM can be used for much more than measuring images, however. One of the unique advantages of SPM techniques is the highly accurate positioning of the probe on or close to the sample surface. This has become an enabling technology for the measurement and manipulation of samples on the nanoscale. AFM's other key advantages are its very high sensitivity, and the fact that the smaller the instrument, the more sensitive it can be. This is the opposite of all previous tools, and means that AFM integration with other techniques is very simple.

Chapter 2

AFM instrumentation

In theory an AFM is a relatively simple instrument. However, constructing an AFM with nanometre-scale resolution requires a considerable amount of sophisticated engineering. The main components of an AFM are the microscope stage itself, control electronics and a computer. The microscope stage contains the scanner (the mechanism for moving the AFM tip relative to the sample), sample holder and a force sensor, to hold and monitor the AFM tip. The stage usually also includes an integrated optical microscope to view the sample and tip. Often, the stage is supported on a vibration isolation platform which reduces noise and increases the resolution obtainable. The control electronics usually takes the form of a large box interfaced to both the microscope stage and the computer. The electronics are used to generate the signals used to drive the scanner and any other motorized components in the microscope stage. They also digitize the signals coming from the AFM so that they can be displayed and recorded by the computer. The feedback between the signals coming out and going back into the AFM stage is handled by the control electronics, according to parameters set via the computer. Software in the computer is used by the operator to acquire and display AFM images. The user operates the software program, and the relevant acquisition parameters are passed onto the control electronics box. The computer usually also contains a separate program to process and analyse the images obtained. A photograph of a typical AFM illustrating these components is shown in Figure 2.1.

2.1 Basic concepts in AFM instrumentation

The three basic concepts that one must be familiar with in order to understand the operation of an AFM are piezoelectric transducers (in AFM, often known as piezoelectric scanners), force transducers (force sensors), and feedback control. Basically, the piezoelectric transducer moves the tip over the sample surface, the force transducer senses the force between the tip and the surface, and the feedback control feeds the signal from the force transducer back in to the piezoelectric, to maintain a fixed force between the tip and the sample.

2.1.1 Piezoelectric transducers

Piezoelectric materials are electromechanical transducers that convert electrical potential into mechanical motion. In other applications, they may also be used in the opposite sense, i.e. if a change is caused in the material's dimensions they will generate an electrical potential. Piezoelectric materials are naturally occurring and may be crystalline, amorphous or even polymeric, although the materials used for AFM are generally synthetic ceramic materials. When a potential is applied across two opposite sides of the piezoelectric device, it changes geometry. The magnitude of the dimensional change depends on the

Fig. 2.1. Photo of a desktop AFM illustrating the major components. They are the microscope stage, computer, electronic controller, computer monitor, and optical microscope monitor. The trackball is used for moving the sample stage in the *X-Y* axis. Resolution can usually be improved by placing the microscope stage on a vibration isolation table.

material, the geometry of the device, and the magnitude of the applied voltage. This is illustrated schematically in Figure 2.2.

Typically, the expansion coefficient for a single piezoelectric device is on the order of 0.1 nm per applied volt. Thus, if the voltage used to excite the piezomaterial is 2 volts, then the material will expand approximately 0.2 nm, or approximately the diameter of a single atom. It is the ability to accurately control such tiny movement that makes piezoelectric materials so useful for AFM. Thus, piezoelectric materials are used for controlling the motion of the probe as it is scanned across the sample surface. Piezo-electrics are available in a variety of sizes and shapes, and are generally used in more complex geometries than depicted in Figure 2.2, so that they can scan the tip in multiple

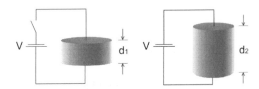

Fig. 2.2. A piezoelectric disk will expand radially ($d_2 > d_1$) when a voltage potential is applied to the top and bottom electrodes. The disk will change shape such that volume is preserved.

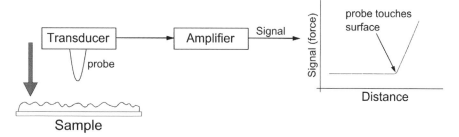

Fig. 2.3. Scheme of force transducer operation. The function of the transducer is to measure the force between the AFM probe tip and the sample surface.

directions across the sample surface. Section 2.2.1 describes in greater detail how piezo-electric materials are configured to scan a probe in three dimensions.

2.1.2   Force transducers

The force between an AFM probe and a surface is measured with a force transducer. As illustrated in Figure 2.3, when the probe comes into contact with the surface, the voltage output from the transducer increases. It is important that the output of the transducer be monotonic and increases as a greater force is applied between the probe and surface. Force transducers may be constructed that measure forces as low as 10 piconewtons between a probe and a surface. Typically, the force transducer in an AFM is a cantilever with integrated tip (the probe), and an optical lever; however, there are several types of force sensors that may be used in an AFM (these are described in Section 2.2.2).

2.1.3   Feedback control

The reason an AFM is more sensitive than a stylus profiler that simply drags a tip over the sample surface, is that feedback control is used to maintain a set force between the probe and the sample. As illustrated in Figure 2.4, the control electronics take the signal from the force transducers, and use it to drive the piezoelectrics so as to maintain the probe–sample distance, and thus the interaction force at a set level. Thus, if the probe registers an increase in force (for instance, while scanning the tip encounters a particle on the surface), the feedback control causes the piezoelectrics to move the probe away from the surface. Conversely, if the force transducer registers a decrease in force, the probe is moved towards the surface. Section 2.3.2 has a more detailed discussion of feedback control methodologies in AFM.

2.1.4   AFM block diagram and requirements

In general terms the design of an AFM is as shown in Figure 2.5. The force transducer measures the force between the probe and surface; the feedback controller keeps the force constant by controlling the expansion of the z piezoelectric transducer. Maintaining the tip–sample force at a set value effectively also maintains the tip–sample distance fixed. Then, the x-y piezoelectric elements are used to scan the probe across the surface in a raster-like pattern. The amount the z piezoelectric moves up and down to maintain the

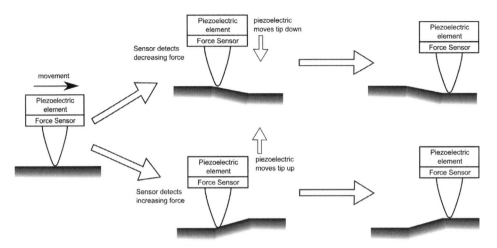

Fig. 2.4. Schematic of feedback control; when the force sensor senses a change in sample height, the piezoelectric moves to maintain the same tip–sample force.

tip–sample distance fixed is assumed to be equal to the sample topography. In this way, by monitoring the voltage applied to the z piezo, a map of the surface shape (a *height image*) is measured.

There are several engineering challenges that must be met to design and construct a successful atomic force microscope. They are:

- A very sharp probe must be constructed so that high-resolution images are measured.
- To get the probe within the scanning range of the surface, a macroscopic translation mechanism must be constructed.
- The force transducer must have a force resolution of 1 nN or less so that the probe is not broken while scanning.
- A feedback controller that permits rapid control so that the probe can follow the topography on the surface must be created.
- An *X-Y-Z* piezoelectric scanner that has linear and calibrated motion must be used.
- A structure that is very rigid must be constructed so that the probe does not vibrate relative to the surface.

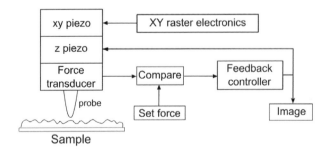

Fig. 2.5. Block diagram of AFM operation.

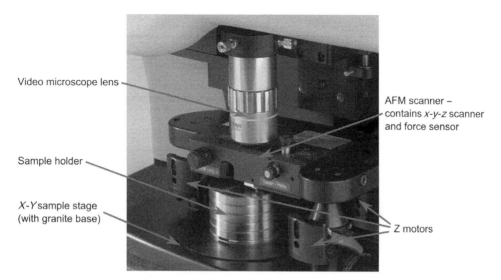

Video microscope lens

AFM scanner –
contains *x-y-z* scanner
and force sensor

Sample holder

X-Y sample stage
(with granite base)

Z motors

Fig. 2.6. Photo of an AFM stage, with components highlighted.

- A high-speed computer that can display the images in real time as they are collected must be used.
- A stage that allows rapid exchange of the probe used for scanning must be created.

The ways in which these challenges are overcome are discussed in the following sections of this chapter.

2.2   The AFM stage

The AFM stage is the heart of the instrument; Figure 2.6 shows an AFM stage and highlights the major components. There must be probe and sample holders. There is a coarse approach mechanism, the Z motor, which can move the AFM scanner towards the sample. There is also an *X-Y* positioning stage which is not required but is useful for positioning the feature for imaging under the probe. To help with this, there is usually an optical microscope for viewing the probe and surface.

A mechanical structure is required to support the AFM scanner and other components. In the construction of the stage it is important that the mechanical loop, which contains all the mechanical components between the probe and surface, be very rigid. If the mechanical loop is not rigid, then the probe will vibrate relative to the sample and introduce unwanted noise into the images.

In general, if the microscope stage is smaller, it will be less susceptible to external vibrations. Creating a rigid mechanical loop becomes more difficult the larger the sample size is. The highest resolution AFMs tend to be very small so that the mechanical loop is rigid, and the microscope stage is not susceptible to external environmental vibrations (or noise).

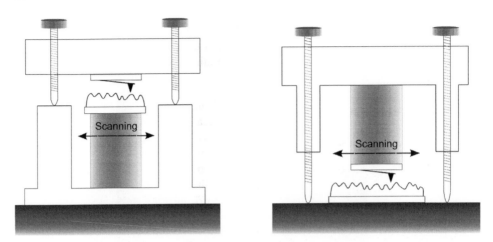

Fig. 2.7. The difference between sample-scanning (left) and probe-scanning (right) microscopes. In a sample scanning AFM the sample is mounted on an x-y-z scanner and the force sensor remains fixed. In the probe scanning AFM the sample remains fixed and the probe is scanned. The advantage of a probe scanning AFM is that it can scan larger samples.

In this book, the motion control mechanisms of the AFM stage capable of moving several millimetres or greater (the coarse movement controls) are designated X, Y and Z. The motion control mechanisms that are used for moving small distances (the x, y and z scanners) are designated x, y and z.

The design of all AFM instruments can be divided into two different configurations as illustrated in Figure 2.7. In the first configuration (left) the sample is scanned and the force sensor is held in one place. In the second configuration, the sample is held fixed and the probe is scanned. In general all AFMs can be divided into such sample-scanning or probe-scanning microscopes. For sample-scanning AFMs, the mass of the sample is included in the feedback loop, reducing the size of sample that may be probed, as well as practical limits on the sample's dimensions. The advantage of the probe scanning (also known as tip-scanning) microscope is that it can be used on any size of sample. In addition, because there is nothing underneath the scanning probe except the sample, it is simple to add accessories to this type of microscope. For example, a liquid cell is easier to use with a probe scanning microscope, and they are easier to integrate with additional optical options, for example to irradiate the sample from the side while scanning, or to mount the entire AFM in an optical microscope. However, the construction of a probe scanning microscope is much more difficult, as the whole tip–optical-lever assembly must be moved while scanning, and care must be taken not to introduce further vibrations from the scanning mechanism into the probe. A sample scanning AFM design is rather simpler, but somewhat limits sample size.

2.2.1   x-y-z scanners

Typically, the scanners used for moving the probe relative to the sample in an AFM are constructed from piezoelectric materials. This is because such piezoelectric materials are readily available, easily fabricated in desirable shapes, and cost effective. However, scanners for AFM may be constructed from other types of electromechanical devices such as flexure stages [27, 28], voice coils [29], etc. All that is important is that the electromechanical device must have very accurate positioning.

2.2.1.1 Piezoelectric scanners

The most common types of piezoelectric materials in use for AFM scanners are constructed from amorphous lead barium titanate, $PdBaTiO_3$ or lead zirconate titanate, $Pb[Zr_xTi_{1-x}]O_3$, $0 < x < 1$ (usually abbreviated as PZT). The ceramics may be 'hard' or 'soft', depending on the formulation. This affects how much they can expand, versus the applied voltage, as well as the linearity of the relationship between applied voltage and expansion. Hard ceramics have smaller coefficients of expansion, but are more linear. Soft ceramic formulations have more non-linearities and have greater expansion coefficients. After fabrication, piezoelectric ceramics are polarized. Polarization may be lost by elevating the piezos to a temperature above their critical temperature or by applying too high a voltage.

Electronically, piezos act as capacitors and store charges on their surface. Capacitances of ceramics may be as large as 100 microfarads. Once a charge is placed on the piezoceramic, the piezoceramic will stay charged until it is dissipated. Electronic circuits used for driving the piezoceramics in an AFM must be designed to drive large capacitive loads.

All piezoceramics have a natural resonance frequency that depends on the size and shape of the ceramic. Below the resonance frequency, the ceramic will follow an oscillating frequency, at resonance there is a $90°$ phase change, and above resonance there is a $180°$ phase change. To a great extent, the resonance frequencies of the piezoelectric ceramics limit the scan rates of atomic force microscopes. As a rule of thumb, the higher the resonant frequency of the scanner, the faster you can scan.

Piezoelectric materials can be fabricated in several shapes such that they have more or less motion. As an example, a disk, as illustrated in Figure 2.2, gets longer and narrower when a voltage is applied. The piezoelectric ceramic changes geometry such that the volume is preserved during extension. Another configuration for a piezoelectric ceramic is a tube, with electrodes on the inside and outside. This configuration gives a lot of motion, and is very rigid. Another configuration is the bimorph, constructed from two thin slabs of piezomaterial that are polarized in opposite directions. When a voltage is applied the ceramic expands in a parabolic fashion. The motions of these geometries, along with the equations of motion are illustrated in Figure 2.8.

Ideally, the piezoelectric ceramics would expand and contract in direct proportion to the driving voltage. Unfortunately, this is not the case, and all piezoelectric materials show non-linear behaviour. They show two primary non-ideal behaviours, hysteresis and creep [30]. Hysteresis, derived from the word history, causes the ceramic to tend to maintain the shape that it was in previously. As the ceramic is expanding, there is a negative shaped non-linearity, and as the material is contracting, there is a positive shaped non-linearity. Hysteresis causes a 'bending' distortion in the images obtained, unless corrected. Creep occurs when the ceramic is subjected to a sudden impulse such as a voltage step function. This means that when the piezo is used to move to a different part of the scan range by applying an offset voltage to it, it will tend to continue moving in the same direction as the offset, even after the voltage has stopped changing. Both these effects are illustrated in Figure 2.9. Real examples of the effects of these non-linearities on AFM images can be found in Chapter 6. These non-ideal behaviours must be corrected to avoid such distortions in the AFM images.

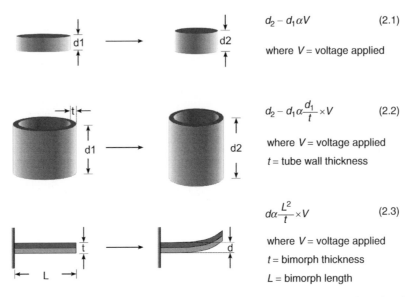

$$d_2 - d_1 \alpha V \qquad (2.1)$$

where $V =$ voltage applied

$$d_2 - d_1 \alpha \frac{d_1}{t} \times V \qquad (2.2)$$

where $V =$ voltage applied

$t =$ tube wall thickness

$$d \alpha \frac{L^2}{t} \times V \qquad (2.3)$$

where $V =$ voltage applied

$t =$ bimorph thickness

$L =$ bimorph length

Fig. 2.8. Typical geometries for piezoelectric elements used in AFM. From top: piezoelectric disk, tube and bimorph scanners.

Correcting the non-ideal behaviours of piezoelectric ceramics is essential for making accurate measurements with an AFM. Due to the different ways the axes are operated – x and y in a raster pattern, z moved by the feedback control – the corrections required are different for the x-y axes and the z axis. Also, hysteresis and creep make it difficult to scan the AFM very quickly, and maintain accuracy. The non-ideal motions of piezoelectric ceramics may be corrected using open-loop or closed-loop methods [31]. The following sections describe the typical methods used to correct for non-linearities in piezoelectric scanners.

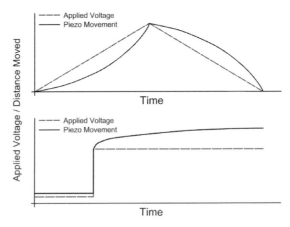

Fig. 2.9. Examples of non-linear behaviour in piezoelectric scanners. Top: hysteresis; when a voltage ramp is applied to the piezo, the response is non-linear. Bottom: creep; after an impulse applied to the piezo, the movement continues in the same direction.

2.2.1.2 x-y *axis correction*

Open-loop techniques require calibration of the AFM scanner to measure the non-linearities. Then the image is corrected using the measured non-linearities. In practical terms, one must measure a very well-known sample, with repetitive patterns to be able to determine the non-linearities in the scanner accurately. Calibration specimens may be bought, or are supplied with instruments for this purpose. See Appendix A for a list of useful materials that may be used in calibrating an AFM. The most commonly used calibration specimens take the form of a lithographically produced silicon grid. Such samples are adequate for calibration in the hundreds of nm to micrometer scale. However, as the non-linearities vary with scan size, further calibration is required for atomic-resolution scanning. Typically, this means scanning a sample with a well-known atomic structure. Once calibrated, the AFM control software will alter the voltage used to excite the ceramic in real time, while scanning, to compensate for the non-ideal behaviour. Alternatively, after an image is measured it may also be 'corrected' by applying a correction function that was previously created. Again, as the calibration factors depend on the scanning conditions, care must be taken to replicate all the scanning parameters exactly, if one is to follow this route. In addition, it is worth remembering that after production, piezoelectric scanners 'relax' slightly, over a long period of time. So, even if a new scanner is perfectly calibrated, after a year or so the person responsible for the instrument should recheck the calibration to maintain accuracy. Procedures to recertify AFM scanners are described in Appendix B. Open-loop techniques are adequate for correcting non-linearity when making measurement with pre-determined scan ranges and speeds. However, open-loop techniques cannot correct for problems associated with creep, and are not really suitable where accuracy is of more importance than high resolution (i.e. metrological applications). In these cases, it is necessary to use external calibration.

An external position sensor can be used in an open-loop or closed-loop design. In the open-loop configurations, the position of the scanner is measured; then the image is corrected after it is measured. In the closed-loop configuration, the motion of the probe is corrected in real time with a feedback electronic circuit. Figure 2.10 shows the use of external sensors in a closed-loop design. Piezoelectric scanners operating with position sensors in a closed-loop scanning configuration are often termed linearized scanners, as it is only in this configuration that their movement is linear.

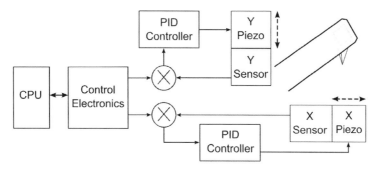

Fig. 2.10. Block diagram for an *x-y* closed-loop scanner configuration.

2.2.1.3   Piezoelectric displacement sensors

Many types of position sensors may be used for correcting the unwanted characteristics in piezoelectric materials. The position sensor must be small in size, stable over long time periods, easily calibrated, have very low noise levels, and be easily integrated into a scanner. Several types of position sensors are available including light-based sensors [31], strain gauges, induction sensors, and capacitance sensors. Optical sensors available include a simple design based on a knife edge attached to the scanner occluding a light beam. [32] The signal from a photodetector is reduced as the knife edge cuts the beam. Other types of light-based motion sensors include using a pinhole above a position sensitive detector and a light lever. Each of these light-based designs requires a high-gain amplifier. The primary advantage of the light-based position sensors is that the parts required for construction are relatively inexpensive. There are many disadvantages however, including the fact that the sensor is not inherently calibrated, misalignments of the light source cause problems, high noise, and the requirement for a high-gain amplifier. The light sources also can cause thermal drift in the AFM scan head. Interferometers may also be used for this function [28, 33], but they tend to be rather bulky and difficult to integrate into the AFM head. Capacitance-based motion sensors are simple devices that measure the capacitance between two plates which depends on the distance, d (Figure 2.11) between the plates, and thus can make a highly sensitive position detector. Capacitance sensors are common primarily because the electronics for capacitance sensors are very sensitive, and

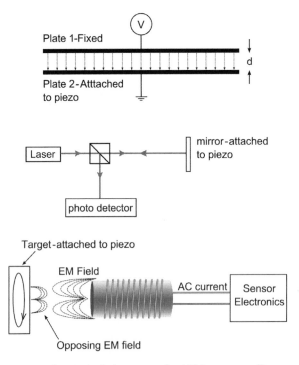

Fig. 2.11. Different approaches to include sensors in AFM scanners. Top: a capacitive sensor, middle, an interferometer-based sensor; bottom: an inductive sensor.

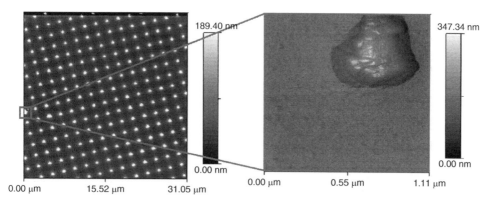

189.40 nm

0.00 nm

347.34 nm

0.00 nm

0.00 μm   15.52 μm   31.05 μm

0.00 μm   0.55 μm   1.11 μm

Fig. 2.12. Zoom to feature example. In this case, with a linearized *x*-*y* scanner, selecting the small feature allowed an immediate zoom to the correct region. Non-linearized scanners cannot accurately zoom to small regions of the scan range.

they are also compact, and so simple to integrate into the AFM. Temperature-based strain gauges may be used. Strain gauges can be attached directly to the piezoelectric material or they may be attached to a structure which flexes when the piezoceramic expands.

Induction sensors are far more suitable for measuring the displacement of the piezo-electrics in SPM scanners compared to optical sensors. Inductive scanners are constructed from a coil through which an AC current flows generating a pulsating electromagnetic field surrounding the coil. Placing the coil a nominal distance from an electrically conductive 'target' induces a current to flow on the target. The induced current produces a secondary magnetic field that reduces the intensity of the original field. The strength of the electromagnetic coupling between the sensor and target depends upon the gap between them, so that the sensor can measure the movement of the scanner. In comparison to optical sensors, induction sensors are small and easily integrated into the AFM scan head in all three axes, have low noise, are stable, and not subject to drift, and only require calibration once at the factory. Other position sensors for AFM piezos based on the interaction of magnetic fields have also been used, and can give very low noise levels. These differ from the induction sensor described here in the geometry of the magnetic field-producing elements. Figure 2.11 illustrates the mode of operation of a number of position sensors commonly used for measurement of piezoelectric element movement.

One of the advantages of closed-loop scan correction is that the scanner movement can be fully calibrated. Such calibrations can give very precise and accurate motion control. However, the calibration procedure can be very time-consuming. Some of the motion sensors, such as the optical-based sensor, are non-linear and require regular recalibration. Other types, such as the inductive and capacitative sensors are reasonably linear and rarely require calibration.

Zoom to feature
One of the problems with AFM scanners with open loop or no scan correction is that it can be difficult to zoom from a large scan range to a specific smaller scan range (zoom to feature, Figure 2.12). Without scan linearization, zooming from a large scan range to a smaller range requires several scans, if one is to be sure not to lose the feature of interest.

However, with the scan calibration sensors operating in a closed-loop configuration, zooming to a specific scan location requires no intermediate scans.

z axis measurement

Correction of hysteresis and creep in the z axis is different from the correction in the xy axis. This is because the xy axis motions are predetermined and the z axis motion is non-deterministic, and depends on the surface topography of the sample being scanned. It is not possible to predict the surface topography, so closed-loop methods will not work. Therefore, AFMs with z calibration sensors use an open-loop configuration for measuring heights. In a z-sensored AFM, when accurate height data is required, the z-sensor signal is used instead of the z voltage to directly measure the height signal. Typically, the AFM software will allow the user to use either the z voltage signal (which has lower noise, and is thus more precise), or the z sensor signal (which is more accurate).

2.2.1.4 Three-dimensional x-y-z scanner configurations

Piezoelectric ceramics must be configured so that they can move the probe, or sample, in the X, Y and Z axes. There are a few standard configurations that are used in AFM instruments. They are the tripod, the tube, and flexures (see Figure 2.13). Each of these designs may be configured for more or less motion, depending on the application for which the scanner is being used. It is also possible to create scanners that use a combination of any of the three basic designs. Currently, the tube scanner is the most widely used, and is the scanner configuration present in >75% of AFMs in use. This type of scanner is so widely used because it is very compact, allows very precise movements especially at small scan ranges, but mainly because it is simple to fabricate. It is also particularly convenient to engineer a probe-scanning AFM with a tube scanner, because there is a clear optical path down the centre of the tube. However, it has some disadvantages; tube scanners, due to their geometry are subject to a lot of non-linearity, particularly bow (an example of the effect of scanner bow is shown in Section 6.2), when using the full range of the scanner.

2.2.1.5 Scanners for fast AFM

In order to develop an AFM that is able to scan much faster than normal, the scanner must be able to overcome the limitation of the traditional scanners, which is their low first resonant frequency. A scanner with a higher resonant frequency will allow faster scanning without the scanner going into resonance. Ando *et al.* have made significant progress in this direction [3, 4, 34]. For example, a fast scanner has been constructed from piezoelectric stacks, to achieve a high resonant frequency of 240 kHz versus 15 kHz for a typical tube scanner [34]. An alternative technique is to use resonant scanners [35, 36]. This means a very high scan rate can be used, but the scan rate is fixed. Typically these are constructed from high resonant frequency flexure scanners, or can also be constructed with tuning fork arrangements, although these are somewhat impractical for large samples [35].

Fast scanning AFM systems have been shown to achieve scanning as fast as 80 ms per frame in intermittent contact mode in liquid [4], or even as fast as 1 ms in contact mode (albeit without full feedback) [35], compared to *ca.* 100 seconds for a normal AFM. However, in order to scan samples with significant topography, the greatest challenge is to create a z-axis positioner whose response is fast enough to react to rapid changes in the sample height, due to extremely fast x-y scanning over the sample.

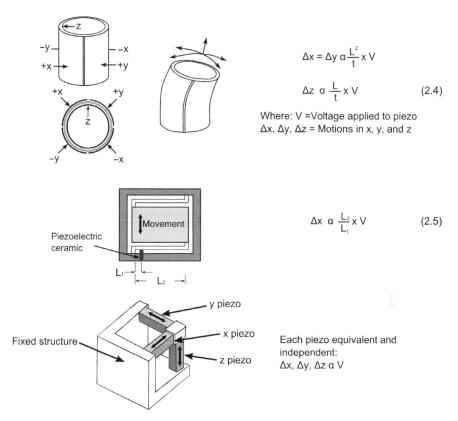

$$\Delta x = \Delta y \; \alpha \frac{L^2}{t} \times V$$

$$\Delta z \; \alpha \; \frac{L}{t} \times V \qquad (2.4)$$

Where: V = Voltage applied to piezo
$\Delta x, \Delta y, \Delta z$ = Motions in x, y, and z

$$\Delta x \; \alpha \; \frac{L_2}{L_1} \times V \qquad (2.5)$$

Each piezo equivalent and
independent:
$\Delta x, \Delta y, \Delta z \; \alpha \; V$

Fig. 2.13. Configurations of common AFM scanners. Top: a tube scanner is configured so that it moves in the *x-y-z* axes. Four electrodes on the outside are used for the *x-y* axis motion, and the inner electrode is used for the *z* axis motion. Middle: a flexure scanner operates by pushing on a flexure with a piezoelectric which then causes the stage to move. There is a gain in the motion given by the ratio of L_2/L_1. A one-dimensional flexure is shown for clarity, typically flexure scanners are set-up to scan in the *x-y* axes. Bottom: the simplest three-dimensional scanner, the tripod scanner.

2.2.2   Force sensors

The force sensor in an AFM must be able to measure very low forces. This is because, for a very sharp probe to be used, a low applied force is required so that the pressure (force/area) can be low enough so that the probe is not broken.

A number of different force sensors have been tested and demonstrated to work with an AFM. Some of these force sensor designs are illustrated in Figure 2.14. The use of an optical lever (sometimes known as a light lever), used routinely for measuring minute motions in scientific instrumentation, was first demonstrated in an AFM in 1988 [23]. With the advent of microfabricated cantilevers the optical lever AFM became the most widely used design for the force sensor in an AFM, and today, nearly all AFMs employ optical lever force sensors.

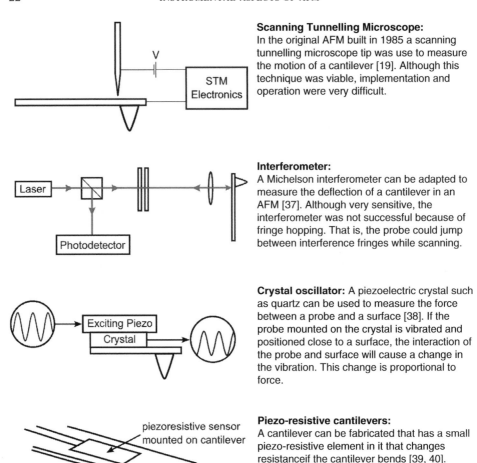

Scanning Tunnelling Microscope:
In the original AFM built in 1985 a scanning tunnelling microscope tip was use to measure the motion of a cantilever [19]. Although this technique was viable, implementation and operation were very difficult.

Interferometer:
A Michelson interferometer can be adapted to measure the deflection of a cantilever in an AFM [37]. Although very sensitive, the interferometer was not successful because of fringe hopping. That is, the probe could jump between interference fringes while scanning.

Crystal oscillator: A piezoelectric crystal such as quartz can be used to measure the force between a probe and a surface [38]. If the probe mounted on the crystal is vibrated and positioned close to a surface, the interaction of the probe and surface will cause a change in the vibration. This change is proportional to force.

Piezo-resistive cantilevers:
A cantilever can be fabricated that has a small piezo-resistive element in it that changes resistanceif the cantilever bends [39, 40]. This type of sensor is viable, but very difficult to manufacture in appropriate quantities.

Fig. 2.14. Different force sensors employed in AFM designs.

The principle of the optical lever is shown in Figure 2.15. The lever consists of a laser focused to a spot on the back of a reflective cantilever; the beam is then reflected onto a split photodetector, which measures the position of the laser spot. In an analogous way to a mechanical lever, the optical lever magnifies a small movement of the cantilever, to create a large movement at the photodiode. The chief advantage of this system is that it is highly sensitive to very small movements of the cantilever, and it is quite simple to build [23, 24, 41].

2.2.2.1 Optical lever sensors

The design for an optical lever AFM sensor is illustrated in Figure 2.15. A laser beam is reflected by the back side of a reflective cantilever onto a four-segment photodetector. If a probe, mounted on the front side of the cantilever, interacts with the surface the reflected light path will change. The force is then measured by monitoring the change in light detected by the four quadrants of the photodetector.

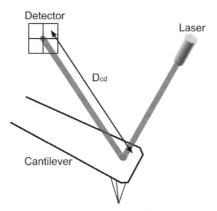

Fig. 2.15. Schematic diagram of the optical lever sensor. In an optical lever, as the end of the cantilever bends the position of the laser spot on the detector changes. As the cantilever–detector distance D_{cd} is large, a small movement of the cantilever causes a large change in the laser spot position at the detector.

The cantilever in the optical lever AFM is typically fabricated with a MEMS process. The cantilevers are small, generally between 50 and 300 microns long, 20–60 microns wide, and between 0.2 and 1 micron thick. Section 2.5 has a more detailed discussion of the cantilevers and probes used in an AFM. The optical lever AFM force sensor requires alignment each time the probe is changed. Typically, alignment is accomplished by first positioning the laser beam onto the cantilever, and then confirming that the light is reflected onto the centre of the photodetector by looking at the photodetector signal. This alignment procedure is rather time-consuming, and is not always fully reproducible; small changes in the laser alignment can affect the force-sensitivity of the system. The alignment procedure is one of the disadvantages of the optical lever system. A procedure for optical alignment is given in Section 4.2. The laser can also give rise to image artefacts as shown in Section 6.6. In the ideal optical AFM design, the probe would have a 90° angle with respect to the surface. Practically, however, this is not possible because of the constraint of the mechanism that holds the probe in place. This requires that there be an angle between the probe/cantilever and the surface, to ensure that only the tip of the probe touches the sample. This angle is usually between 5° and 15°. Such angles can also cause artefacts in the images. Some probes are available with a counter-angle built into the geometry, i.e. the tip is mounted onto the cantilever at *ca.* 12° so that it can approach the sample at an angle close to the perpendicular. The optical lever sensor is by far the most widely used force sensor for AFMs. The following sections cover the design and implementation of optical lever force sensors.

2.2.2.2 *Integrating optical lever force sensors and scanners*

The first AFM designs scanned the sample and kept the probe stationary. This sample-scanning design is optimal for only limited types of sample. To create tip-scanning AFMs it is necessary to design AFM scanners where the *x-y-z* scanner is integrated with the

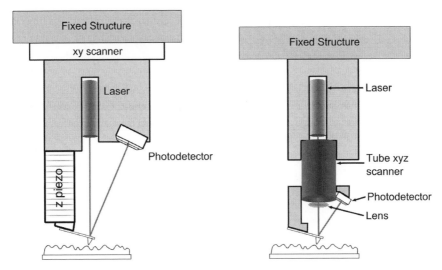

Fig. 2.16. Designs for tip-scanning AFMS with optical lever sensors. Left: the laser is scanned with the cantilever. Right: the laser is fixed and the cantilever is scanned, a lens keeps the laser light focused on the cantilever.

optical lever AFM force sensor. The simplest approach to integrating the x-y-z scanner would be to mount the optical lever sensor at the end of the scanner. This is not feasible because the z piezo is not responsive enough to move the entire light lever up and down as the probe is scanned across the surface. Such an AFM would be too slow to be practical. Two methods are employed for creating a combined optical lever AFM scanner with an x-y-z scanner.

In the first configuration, illustrated schematically on the left of Figure 2.16, the laser and photodetector are scanned in the X-Y axis, and the probe is mounted at the end of the Z piezoelectric. In this design the z piezo is part of the optical lever optics. This means that as the probe is moved up and down in the Z direction the light path changes. However, it can be shown geometrically that the Z motion of the cantilever has a minimal effect on the operation of the AFM optical lever AFM sensor. In this design, commonly the x-y scanner would be a flexure scanner, and the z scanner a simple piezo stack. Also illustrated in Figure 2.16 is the other approach that is commonly used. The laser is held fixed and a lens is used to focus the laser light onto the scanning cantilever. As the lens moves back and forth in the X-Y plane, the laser light stays focused on the cantilever. The photodetector must be then mounted on the x-y translator.

2.2.3 Coarse Z movement – probe–sample approach

One of the major challenges in AFM design is making a motion control system that permits the approach of the probe to the surface before scanning. This must be done such that the probe does not crash into the surface and break. An analogous engineering challenge would be to fly from the earth to the moon in 60 seconds and stop 38 meters from the surface without overshooting or crashing.

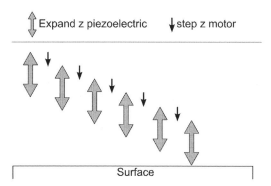

Fig. 2.17. 'Woodpecker' probe approach method. The surface is approached by alternately expanding the piezoelectric element, and stepping the z motor. This avoids uncontrolled contact between the probe and the sample. As soon as the surface is encountered, the feedback system is turned on.

In the AFM stage there are two separate motion generation mechanisms in the Z axis. The first is a stepper-motor-driven mechanism with a dynamic range of a centimetre and a resolution of a few microns. The stepper motor is driven either by a linear bearing or an 80 turn per inch screw. The second motion generation mechanism in the Z axis is the z piezoelectric element in the AFM scanner. The z piezo typically has a dynamic range of about 10 microns or less and a resolution of less than 0.5 nm. While stepper motors have the range and speed to approach the surface from a great distance in a short time, they have neither the resolution nor fast response time to put the tip into feedback safely. On the other hand, the piezo driver is sensitive enough to safely go into feedback, but can only move short distances.

Typically, probe approach is achieved with a 'woodpecker' method, (shown in Figure 2.17). In this method, the stepper motor is stepped a small increment, say 1 micron. Then the z piezoelectric ceramic is extended 5 microns to see if the surface is detected. The z piezo is then retracted, the stepper motor extends one more micron, so on and so on. A key component here is that when the probe encounters the surface, the feedback is turned on immediately. In this way, the AFM can approach the surface from several hundreds of microns, without risk of crashing the tip.

There are two primary mechanisms that may be used for the Z motion control, as shown in Figure 2.18. In the first, three lead screws are used together with a kinematic mount. All three screws can be turned simultaneously or a single screw may be turned. If only one of the screws is turned, there is a reduction of motion at the centre of the three screws. This geometric reduction in motion can be used to get very precise motion. For automated tip approach, one or all of the lead screws is attached to a motor. In the second method, a linear bearing is used to drive the AFM scanner towards the sample. The linear bearing must be very rigid to avoid unwanted vibrations.

2.2.4   Coarse X-Y movement

Most AFMs include an X-Y position stage for moving the sample relative to the probe. The stage may be manual or automated with motors. The primary function of the X-Y stage is

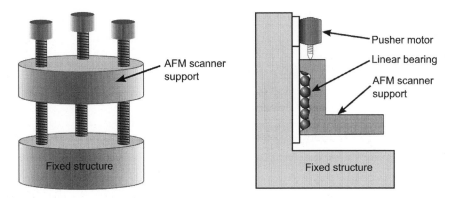

Fig. 2.18. Configurations used for coarse Z approach mechanism. Left: on AFMs designed for small samples, a kinematic mount is typically used. One or all of the threaded screws are usually motorized for an automated probe approach. Right: a linear bearing could also be employed to move the AFM head in the Z axis.

for locating features on a surface for scanning with the AFM. The resolution of the X-Y stage is usually less than 1/10 the range of the x-y scanner that moves the probe. There are two possible configurations for the X-Y stage. In the first, the sample sits on top of an x and y crossed roller bearing. In the second, the sample is mounted to a block that is directly on the base of the microscope. Typically the base is made from granite. The metal block is then pushed around with the X-Y motors. The advantage of the second design is that there is less chance of the X-Y stage introducing noise into the AFM mechanical loop. In both cases, the mechanisms must be highly wear-resistant, as any vibrations will compromise the mechanical loop of the AFM head.

2.2.5 Optical or inspection microscope

Like the X-Y stage, the microscope optic is not an essential feature for an AFM stage. The optic is generally used for finding the region for scanning. Also, the optical scope can be helpful in positioning the laser light on the cantilever in the optical lever AFM force sensor. The optical microscope in an AFM can also be helpful for probe approach.

There are three optical microscope viewing designs that may be used in an AFM stage, illustrated in Figure 2.19. The 90° top down design is optimal for applications when high-resolution optical microscope imaging is mandatory. The 45° design is particularly helpful for probe approach and is used when high-resolution optical imaging is not required. The 90° bottom view design is typically used with an inverted optical micro-scope for biological applications. In this case, it is particularly useful to use a probe-scanning design, usually with a tube piezo or other scanner that can have a hole in the centre. Optical access is then unimpeded, and the lack of any AFM components below the sample reduces the chance of instrument damage from buffer solution leaks or temperature effects; AFM scanners are generally incompatible with water, or great temperature variations. The integration of high quality inverted optical/fluorescence/confocal microscopes with AFM is very useful in a range of biological applications, see Section 7.3, and AFMs have been designed for integration with such microscopes since the 1990s [42, 43].

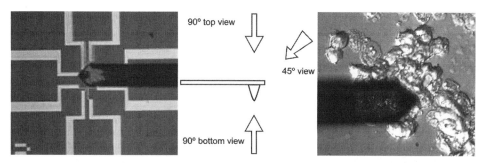

Fig. 2.19. Left: video camera image of the cantilever and sample in an AFM (90° top view). The red 'spot' is from the laser that is used in the optical lever force sensor. With scanning ranges greater than 1 μm, it is possible to see the AFM cantilever move in the video microscope image. Middle: the three possible viewing positions of an optical microscope in an AFM. Right: image in an AFM with 90° bottom view; note the laser light (purple in this case) can be seen through the cantilever, which is seen through the sample (cells on a glass slide). (A colour version of this illustration can be found in the plate section.)

2.2.6 Mechanical loop

The greatest factor that affects the vertical resolution or noise floor of an AFM is the rigidity of the mechanical loop. The mechanical loop is comprised of all the mechanical elements between the sample surface and the probe, as illustrated in Figure 2.20. If this loop is not rigid, then the probe can vibrate in an uncontrolled manner relative to the sample, and noise is introduced into images. It is typically easier to make the mechanical loop very rigid by making the microscope very small. Because of this, in practice the highest resolution AFM instruments are very small. It also means that it is very difficult to make AFM stages for larger samples such as silicon wafers or optical disks that have very high vertical resolutions.

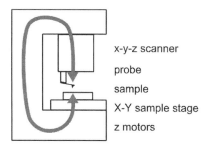

x-y-z scanner

probe

sample

X-Y sample stage

z motors

Fig. 2.20. The mechanical loop in an AFM includes all of the structural elements that are required to hold the probe at a fixed distance from the sample. This includes the x-y-z scanner, *X-Y* sample stage, *Z* motor and the probe.

2.3   AFM electronics

Most of the electronics in an AFM are resident in a separate cabinet from the stage and the computer. The functions in the electronic controller may be constructed with digital signal

processing (DSP) chips or analogue electronics. This section does not discuss the implementation, but describes the block functions in the controller. The primary function of the
electronics in an AFM is to:

(a) Generate scanning signals for the x-y piezoelectrics.
(b) Take an input signal from the force sensor and then generate the control signal for
    the Z piezo.
(c) Output control signals for X-Y-Z stepper motors.
(d) Generate signals for oscillating the probe and measuring phase or amplitude when
    an oscillating mode is used for scanning.
(e) Collect signals for display by the computer.

As mentioned above, these functions may be implemented with either digital or analogue
electronics. In the digital approach, see Figure 2.21, all signals from the stage are digitized,
and a DSP chip takes care of all of the feedback control calculations. Also, the DSP chip
generates the x-y raster scan functions. The advantage of analogue electronics is that they
are typically less noisy. This will generally lead to a lower noise floor of the instrument,
and thus may enable acquisition of higher resolution images. Because the functionality of
a DSP chip is created by a software program, the DSP approach gives more flexibility and
can be changed very rapidly. Instruments with digital electronics might, for example,
allow simple software 'upgrades' to enable new features or acquisition of more data
channels simultaneously. The following sections are a detailed description of the functions
shown in Figure 2.21.

2.3.1   x-y signal generation

The x-y signal generator create a series of voltage ramps that drive the x and y piezoelectric
elements in the AFM, as illustrated in Figure 2.22. The scan range is established by
adjusting the minimum and maximum voltage. The position of the scan is established by
offsetting the voltages to the ceramic. Finally, the scan orientation is rotated by changing
the phase between the signals. It can be seen from Figure 2.22, that the forward and reverse
scan lines do not cover exactly the same topography. However, it is usually assumed that
they are equivalent, and generally, there is no appreciable difference between the two. In
general it is best if the drive signals do not have sharp edges at the turning point. Sharp
edges can excite resonances in the piezoelectric ceramics, and cause them to vibrate. Such
vibrations create unwanted artefacts and 'ringing' in the images. Higher speed scanning
with an AFM in particular is almost always done using rounded signals such as sinc waves
to drive the piezoelectric ceramics. Furthermore, even with slow speed AFM when using
straight-edged signals such as shown in Figure 2.22, the response of the scanner is not
linear at the turnaround points. To overcome this some 'overscan' is typically included in
the scanning, such that only the linear response part of the data is recorded. For example,
to scan a 10 μm area, the instrument might really move 12 μm in the slow scan direction,
and discard 1 μm of the data from either end. In this way, the recorded data does not suffer
from edge artefacts.

The maximum scan range of the AFM scanner is established by the mechanical–electrical gain of the piezoceramics and the maximum voltage they can tolerate before
depolarizing. As an example, the piezoceramics may have a gain of 1 μm per volt. If

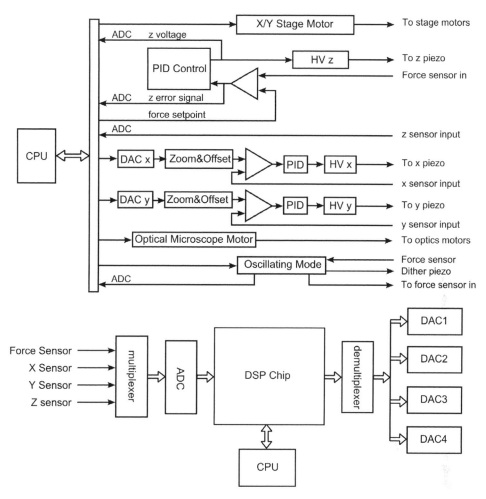

Fig. 2.21. Block diagram of AFM electronics functions. Top: electronics as implemented with analogue electronics; Bottom: as implemented with a high-speed DSP chip.

the maximum potential is 100 volts, then the scan range is 100 microns. The maximum achievable resolution is set by the noise floor of the driving voltage. A noise floor of 1 millivolt would give an X-Y resolution of 1 nanometre in this case.

It is important that the bit noise associated with the X-Y scan generators be less than the analogue noise floor of the electronic controller. For example, if the scan range is 100 μm and the analogue noise floor is less than 1 nm, then the number of bits required is at least 100,000, which is greater than 2 [16] bits. This is significant because most DACs store such data as 16 bit numbers (i.e. they can have no more than 65,536 possible values). Thus, if this is the case, some resolution would be lost when the data was digitized. To overcome this, one option is to use a scale and offset DAC and amplifier if the scanning DAC does not give enough bit resolution. This overcomes the resolution problem because although

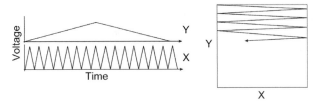

Fig. 2.22. The way the x and y piezoelectric elements are driven by varying potentials. Left: illustration of the signals output for driving the x and y piezoelectrics in the AFM scanner. Right: the motion of the probe in the x and y axis when the piezoelectric ceramics are activated.

with a 100 μm scan range we would like to have 1 nm resolution, we do not require that resolution over the whole range, but rather it's required in small section of the possible range (for example a 512 $\times$ 512 pixel region covering only 1 μm of the range). As an alternative, a DAC with a much higher number of bits may be used. Note the inclusion of circuits for scale and offset in the AFM electronics in Figure 2.21.

2.3.2 Feedback control circuit

In the AFM, the feedback control electronics take an input from the force sensor and compare the signal to a set-point value; the error signal is then sent through a feedback controller. The output of the feedback controller then drives the Z piezoelectric ceramic. The type of feedback control used in AFMs is called a proportional-integral-derivative controller (PID). The equation governing the way this operates is shown in Figure 2.23.

The proportional-integral-derivative controller takes the error signal and processes it as follows: By selecting the appropriate P, I and D terms in Equation 2.6, the probe will 'track' the surface as it is scanned, keeping Z_{err} minimal. The integral term facilitates the probe moving over large surface features and the P and D terms allow the probe to follow the smaller, high-frequency features on a surface. Many AFM instruments actually use a PI controller, as the derivative term is not used, although by convention the controller is still referred to as a PID controller. Here, we follow this convention. The two signals from the feedback loop that are typically digitized to create AFM images are the error signal and the z voltage. The z voltage (converted using the instrument calibration to distance) forms the 'height' or 'topography' image. The use of the error signal is described more thoroughly in Chapters 3 and 4 but most importantly, it is used by the instrument operator to optimize scanning the parameters, including P, I and D values. When the PID parameters are optimized, the error signal image will be minimal. Section 4.2 describes the process for optimizing the PID parameters in an AFM. Implementation of the z feedback loop in an AFM can be made with either analogue or digital electronics. The advantage of digital

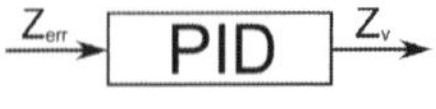

$$Z_{err} \rightarrow \boxed{\text{PID}} \xrightarrow{Z_v} \quad Z_v = P \times V_{err} + I \times \int Z_{err} dt + D \times \frac{dZ_{err}}{dt} \tag{2.6}$$

Fig. 2.23. Proportional-Integral-Derivative (PID) controller operation and equation.

electronics is that they are very flexible and can be configured to do many types of functions. Analogue electronics typically have less noise and have a larger dynamic range. Either approach will typically provide adequate results.

2.3.3 Output of signals for stepper motors

Usually AFM stages have several stepper motors that must be electronically controlled. The stepper motors are typically driven with a series of voltage pulses that are in a specific phase sequence. The functions in the stage that may be controlled with stepper motors include:

- *X-Y* sample translation.
- *Z* motion control (1 to 3 motors). These are for the Z-approach mechanism, which must be coordinated with movements of the *z* piezo scanner (see Section 2.2.3)
- Zoom/focus on video microscope.
- Some instruments allow the user to manually 'step' the *z*-motor a little in order to reposition the scanning position along the *z* piezo.
- Some instruments have focussing/alignment controls for the laser in the optical lever.

Typically, an AFM will have subset of these motorized mechanisms. Simpler AFMs will have fewer of them implemented, as they simply make the AFM more convenient to use. The exception is the Z-approach mechanism which is required for all AFM instruments.

2.3.4 Oscillating signals

For operation of certain AFM modes, it is necessary to mechanically oscillate or vibrate the cantilever and to compare the modulated signal phase or amplitude to the drive oscillation. Section 3.1.2 provides a detailed explanation of the way these modes operate. Feedback control may be implemented such that the phase or amplitude difference to the input signal is kept constant during scanning. Figure 2.24 illustrates the circuit used for mechanical modulation and phase/amplitude detection in the AFM. If the feedback control maintains a constant phase change, then the amplitude may vary while scanning. *Vice versa*, if the amplitude is maintained constant, then the phase may vary while scanning. For this reason, the AFM typically includes A/D converters to capture and display the amplitude and phase signal.

2.3.5 Collecting signals

Many electronic signals associated with the Z axis in the AFM are digitized and may be displayed by the computer. These signals include:

- *z* voltage – The voltage that goes to the *z* piezoelectric ceramic, after the PID controller.
- *z* error signal – This signal is proportional to the output of the light lever photo-detector, also known as the deflection signal.
- Z sensor – The signal from the motion sensor, if present, measures the displacement of the *z* piezoelectric in the AFM scanner.
- Amplitude – The signal from the amplitude demodulator.
- Phase – The signal from the phase demodulator.

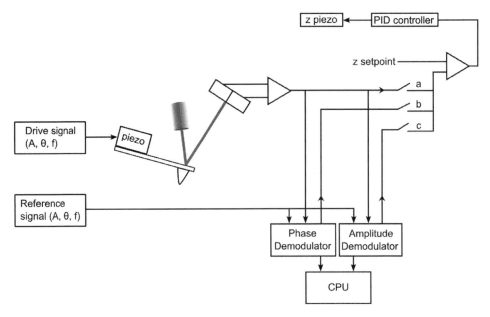

Fig. 2.24. Block diagram of the electronics employed for oscillating mode AFM scanning. The signal used for feedback can be selected by switch a, b or c. Switch a is for DC feedback, b for phase feedback, and c for amplitude feedback.

In an AFM there is typically one or more high-speed analogue to digital converters (ADC). If there is a single ADC, the many analogue signals are passed through a multiplexer into the ADC input (see Figure 2.21). The speed of the A/D converter must be high enough such that at least one data point is converted per pixel.

Note that bit noise, as described in the section about x-y scanning, is also important in the context of the acquisition of the z axis data i.e. the z voltage signal. Although the z piezo range is typically much lower than the x-y range (typically, a large sample AFM scanner might have a z range of 10 μm and an x-y range of 100 μm), the achievable resolution in z is also much greater than in the x-y plane. If we imagine the case above, then with 10 μm z range a 16 bit ADC would limit us to 10,000 angstroms/65,356 bits = 1.4 angstroms per bit. This is much greater than the resolution of a modern AFM, which might be expected to show <0.5 angstrom root-mean-squared (rms) noise in z under typical conditions. This bit noise can significantly degrade results when scanning small features on a very flat surface, or carrying out sensitive force spectroscopy experiments. Typically, to overcome this, the AFM allows a similar 'scale and zoom' solution; the z bit resolution is increased temporarily by only using a part of the z piezo range. This setting is typically applied by the AFM operator, rather than automatically. This is because it should only be used with flat samples and small scans, as it is not advantageous to reduce the z range while scanning rough samples.

2.4 Acquisition software

Typically, a software interface is used for controlling the AFM stage. Functions controlled by software include setting all movement of the X-Y stage to locate the feature for scanning, probe approach to get the probe near/on the surface, selection of scan mode, setting and controlling scan parameters, display of images while scanning, and a capability for measuring force–distance (F/D) curves. Sometimes image processing and analysis are also handled within the same application, but these functions are covered in Chapter 5. Figure 2.25 is a screenshot of a typical AFM scan control window. The following section describes the functions found in the control software.

2.4.1 Display

Visualizing the AFM data in real time is critical to the efficient operation of the AFM. This allows an operator to ensure that they are scanning the correct region of a sample, and facilitates optimizing the scan parameters such as scan rate and PID settings. Typically there are at least two types of display.

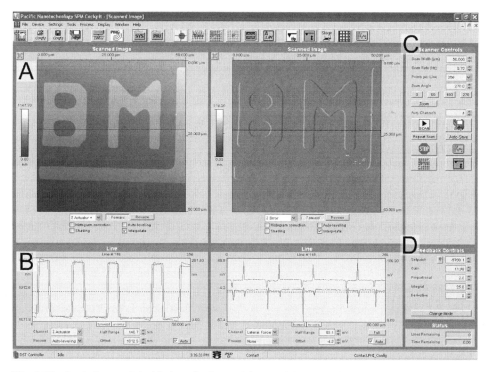

Fig. 2.25. A window such is this is typically used for acquiring AFM images. There are sections for **A**: displaying images; **B**: displaying 2-D profiles, **C**: entering scan parameters; and **D**: entering feedback parameters.

(a) A two-dimensional representation of the image shows the topography of the specimen being scanned. To correct for tilt between the probe and sample, the height image is typically line levelled in real time (see Section 5.1.1 for an explanation of levelling). Without real time line levelling, the image will only show the tilt between the probe and sample. Usually multiple channels can be shown simultaneously.

(b) An 'oscilloscope window' displays a two-dimensional scan line such as the Z signal versus the x axis motion. An oscilloscope window is very helpful for optimizing the scan parameters and ensuring that the probe is tracking the sample's surface. Typically, data collected in the forward and backwards directions data can be overlaid.

(c) As discussed in Section 2.3, there are a number of data channels that may be monitored in the AFM instrument. They include the Z piezo voltage, Z error signal, Z motion sensor, and phase and amplitude signals. The AFM control software will allow one or more of these signals to be displayed on the screen. They may be available as either images, or scan line data or both. In addition to the height data, viewing the z error signal while optimizing acquisition parameters can be particularly helpful.

2.4.2   Stage control

Making an AFM practical to use requires motion control including at least one stepper motor to move the probe relative to the sample in the Z axis. Additional motion control is used for moving the sample in the X-Y axis relative to the probe as well as controlling the zoom and focus of an optical microscope.

(a) Z motion control: Probe approach is a very important function in the AFM (see Section 2.2.3 for the hardware implementation of this and Section 4.2 for precautions for the user on applying z approach). The Z-approach software should be rapid, but it should not allow the probe to touch the surface in an uncontrolled manner. Properly optimized, probe approach takes less than a minute.

The approach software typically has several options for controlling the rate at which the probe moves toward the sample's surface. Software algorithms are also critical for setting the threshold signal levels associated with the probe interacting with the surface. Once the threshold is met, the approach is stopped and the AFM is put into feedback. Properly implemented, a fully automatic approach system prevents the user accidentally inputting a threshold value that would crash the tip into the surface. If there is an automated video microscope, the software algorithm for tip approach can be augmented to shorten the time required for probe approach. This is achieved by focusing the microscope on the probe, then the sample. The relative positions of the probe and sample are compared. Then the Z motors are driven rapidly until the probe is less than 100 microns from the surface.

(b) X-Y motion control: Because the x-y motion using the scanner in the AFM usually has a range of less than 100 microns, an X-Y motion control system is required that is able to move the probe to within a few microns of the features that are to be scanned. An X-Y positioning table driven with stepper motors is often used. Software is then used to move the translation stage. The software typically is activated by mouse control within the software or by a track ball.

Advanced software functions may be added to microscopes with automated X-Y stages. Functions include an ability to measure many images adjacent to each other, and to measure several images on pre-set locations on the sample. It is possible to drive the stage to pre-established locations for inspection applications with registration software.

2.4.3 x-y *scan control*

The exact scan control parameters that are used for scanning a sample depend on the particular application. There are a few variables that must be selected to scan a sample. They are:

(a) Image size: This is the window that is selected for viewing the features on a surface. The image size should be at least as large as the features that are to be visualized. Often a large scan is measured and then the operator 'zooms' in on a feature of interest.

(b) Number of lines in the image: The digital resolution of the image is established with the number of lines selected for the image. For example, if the scan size is 10×10 microns and the number of lines selected is 256 then the digital resolution is 39 nanometres. The number of lines in the image may range from less than one hundred to several thousand. Most AFM software limits the images to square or rectangular dimensions.

(c) Image rotation angle: The image scan angle may be changed with software. Rotation angles between $0°$ and $360°$ can usually be selected. Rotating the image scan axis usually means that the largest scan range cannot be achieved.

(d) Scanning speed: This is typically specified in hertz (i.e. lines scanned per second), and in normal circumstances varies from around 0.5 Hz to 4 Hz. Combining this parameter with the number of lines in the image, gives the time required to collect an image. Along with the feedback parameters discussed in the next section, scanning speed can affect image quality. This is discussed further in Chapter 4. Realistically, the important parameter in terms of imaging quality is distance covered over time rather than frequency of line collection, so scanning speed can also be expressed in micrometres/nanometres per second.

2.4.4 z *control*

Software is required for controlling the feedback control electronics, see Section 2.3. There are two functions that are controlled; the set-point voltage and the PID parameters.

(a) Set-point voltage: This is the voltage that goes into the differential amplifier, so this voltage is compared with the force sensor output voltage and an error signal is generated. The set-point voltage controls the 'relative' force. A calibration of the specific cantilever is required to convert the set-point voltage to a force (see Section 2.5).

(b) PID parameters: These parameters control the 'responsiveness' of the feedback control electronics. These parameters must be adjusted such that the probe tracks the surface while scanning. Sections 4.2–4.3 provide a description of optimization of the feedback control parameters.

2.4.5 *Force–distance curves*

Force–distance (F-D) curves are used to measure the forces experienced by the probe as a function of distance from the surface. In F-D measurements, the probe is moved toward the sample surface to a pre-selected position, and then retracted. The extent of cantilever deflection over the course of this movement is expressed by the Z (deflection) signal which

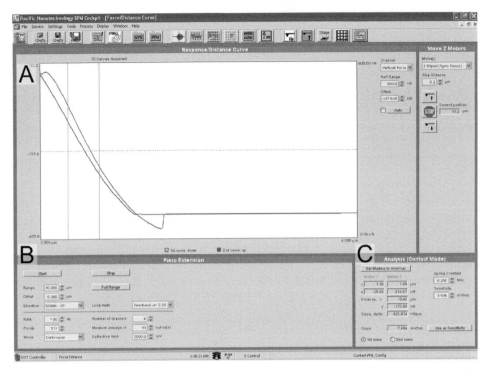

Fig. 2.26. Software windows used for measuring force/distance curves with the AFM. The main window shows the curves acquired (**A**), the window below it allows the entry of acquisition parameters (**B**), and to the right of this window are settings for extracting data from the curves (**C**).

is used to generate a force–distance curve. The end point of the force curve might be defined by the user as a certain distance from the start point, or as a certain value of cantilever deflection. The latter option allows the user to effectively define the maximum force applied. Software for making F-D curves, illustrated in Figure 2.26, has several variable parameters including:

(a) start and end position for probe;
(b) rate of probe approach motion;
(c) number of F/D curves to signal average;
(d) location on image for F/D curve.

Typically, the AFM software will also allow the user to collect a series of F/D curves in a grid pattern over a user defined area of the sample surface, thus enabling measurement of the tip–sample interaction across the sample surface. This facility may be termed layered imaging, volume spectroscopy or force volume imaging.

2.5 AFM cantilevers and probes

An optical lever-based AFM force sensor requires a cantilever with a probe at its end for operation. Typically these are fabricated using MEMS technology and are considered a

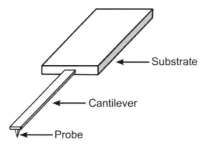

Fig. 2.27. Illustration of an AFM cantilever/probe/substrate created by micromachining of Si or Si_3N_4. All commercially available probes have substrates with the same dimension, for ease of use in different instruments. The probe is sometimes referred to as the tip, and the substrate as the chip. Not to scale.

disposable component of the AFM. In principle, an AFM probe should last forever; however, in practice the probe tip is often blunted when it touches a surface. Changing the probe typically takes only a few minutes. In order to make handling simple, the cantilevers are attached to a cantilever substrate or chip. By industry convention, these are normally *ca.* 3.5×1.6 mm in size, and about 0.5 mm thick, so that probes from different manufacturers can be used in most probe holders built into AFMs. Figure 2.27 shows a cartoon of the design of a typical probe/cantilever/substrate.

The geometry of the probe is critical to the quality of images measured with an AFM. All AFM images are a convolution of probe geometry and surface. As an example, in Figure 2.28, if the probe cannot reach the bottom of a surface pit, or track the sides of a particle, the image will not indicate the correct geometry of the sample. Further details on the problems associated with blunt tips are shown in Section 6.1.

2.5.1 Probe materials

In principle, AFM cantilevers can be fabricated from any material that can be fabricated into a spring-like cantilever. The first AFM cantilevers were fabricated from tungsten wire and

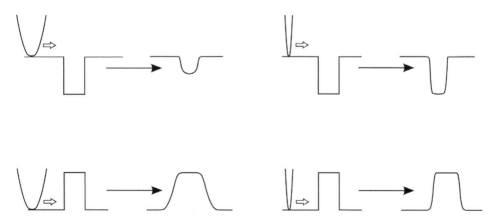

Fig. 2.28. Comparison of image profiles obtained with a dull (left) or a sharp (right) probe on a concave feature (a pit, top) or a convex feature (a step feature, bottom).

Fig. 2.29. Examples of contact and non-contact probes. Left: a typical v-shaped contact-mode cantilever. The whole probe is made from silicon nitride (Si_3N_4), and has an integrated square pyramidal probe tip. Right: a probe designed for oscillating modes such as non-contact AFM. The cantilever is usually rectangular (or a modified rectangle shape like this one). The whole probe is made from silicon, and is much stiffer and more prone to breaking than the contact probe; however it has a sharper tip.

had a probe etched in the silicon at the end. Early in the evolution of AFM it was discovered that the best AFM probes could be constructed from MEMs technology. There are two materials commonly used for AFM cantilevers: silicon nitride (Si_3N_4) and silicon (Si).

Si_3N_4 is used for creating probes that have very low force constants. The thin films used for creating Si_3N_4 probes must have very low stress so the cantilevers don't bend naturally from the stress. Practically, most Si_3N_4 films have some residual stress and in fact, cantilevers made with Si_3N_4 tend to have curvature along their primary axis.

Cantilevers fabricated from silicon tend to have less residual stress than Si_3N_4 and so tend not to suffer from bending. However, the Si probes that are fabricated at the end of the cantilever can be brittle and can be more likely to chip when they contact a surface. Most of the probes used in optical lever-based AFM force sensors are constructed from Si.

2.5.2   Contact versus oscillating mode probes

Cantilevers for optical lever-based AFM can be operated in two basic topography modes; contact (static) mode and oscillating modes, see Section 3.1. The cantilevers used for contact mode have force constants that are typically much less than 1 N/m and are fabricated from either silicon or silicon nitride. On the other hand, oscillating mode cantilevers are usually fabricated from silicon and have force constants that are greater than 10 N/m. Examples are given in Figure 2.29. There are also a large number of other probes available differentiated by differing tip geometries (for example many examples of probes with 'sharpened' and high-aspect-ratio tips are available), cantilever force constants, and coatings. Non-topographic modes are commonly carried out with these speciality cantilevers, see below for more about such cantilevers and Section 3.2 for details of their applications.

The rectangular cantilevers used as AFM force sensors have the same mechanical properties as all cantilevered beams. They have a vertical force constant and resonant frequency given by Equation 2.5. Additionally a cantilever has torsional and lateral bending force constants given by Equations 2.6 and 2.7. The calculations of the corresponding equations for v–shaped levers is considerably more complex, see [44].

$$k_{ver} = w \times \tfrac{E}{4} \times \left(\tfrac{t}{l}\right)^3 \qquad (2.7)$$

$$k_{lat} = t \times \tfrac{E}{4} \times \left(\tfrac{w}{l}\right)^3 \qquad (2.8)$$

$$k_{tor} = w \times \tfrac{G}{3} \times \tfrac{t^3}{l} \times \frac{1}{\left(H + \tfrac{t}{2}\right)^2} \qquad (2.9)$$

where: K_{ver} = vertical force constant
K_{tor} = torsional force constant
K_{lat} = lateral force constant
E = Young's modulus
G = modulus of rigidity
w = cantilever width
l = cantilever length
t = cantilever thickness

2.5.3 Control of tip shape

Horizontal resolution with an AFM can often be improved with sharper probes. There are many techniques available for sharpening AFM probes. Because the sharpness and reproducibility of manufactured probes is one of the limiting factors on the quality of AFM results, new types of probes are under constant development. Examples of this include composite probes such as mixed silicon/silicon nitride probes, and probes terminating in carbon nanotubes. However, it remains a major challenge to produce probes with a reproducible tip radius below 10 nm at a reasonable cost.

- Si_3N_4 probes are sharpened by adding an extra process step that changes the shape of the pit that the Si_3N_4 film is deposited on. However, this technique often gives double tips which can cause substantial artefacts in images. Another option is an additional oxidation/etching process after the probe is manufactured.
- Si probes can be sharpened by chemical etching, ion milling or by adding a carbon nanotube (see Figure 2.30). Each of these techniques can create a sharper probe, but also add to the price of fabricating the probe.

One method for controlling the geometry of the probe on an AFM cantilever is to mount a sphere at the end of the cantilever. The sphere can be mounted directly on a cantilever that does not have a probe. The sphere may also be mounted at the end of a 'plateau probe', or a probe that does not have sharp tip, but instead has a flat plateau at the end of the tip.

Probe damage
The quality of an AFM image is critically dependent on the shape of the probe used for measuring an image. The AFM probe can be severely damaged by tip approach. Handling

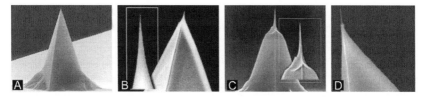

Fig. 2.30. SEM images of different types of sharpened silicon probes. A: a standard silicon oscillating mode probe. B: 'super-sharp silicon' probe sharpened with an electrochemical etch. C: 'high-aspect-ratio' probe sharpened with ion milling. Right: probe modified with carbon nanotube. Images A–C reproduced with kind permission from NanoWorld AG.

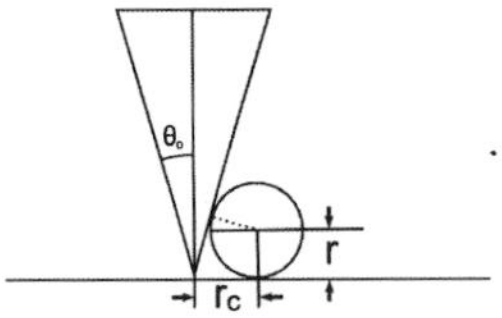

$$\cdot \left(\cos \theta_o + \sqrt{\cos^2 \theta_o + (1 + \sin \theta_o) \left(-1 + \left(\frac{\tan \theta_o}{\cos \theta_o} \right) + \tan^2 \theta_o \right)} \right)$$

(2.10)

Fig. 2.31. Simplified view of probe deconvolution. If the cross-section of the probe geometry is described as an upside down triangle, it is simple to remove the effect of the probe from an image of a nanoparticle. The actual diameter of the nanoparticle is calculated using the equation at the right. However, real probes do not have the shape of an upside down triangle.

the probe incorrectly before it is placed in the microscope can also cause probe damage. For example, if the probe is exposed to high electric fields, the probe tip can be blown off by electrostatic discharge. Finally, the probe tip can get dirty from the packing materials used to hold the probe while shipping.

2.5.4   *Probe shape deconvolution*

If the probe geometry is analytically known, it is possible to remove the probe geometry's effect on an AFM image by a process called deconvolution (see Figure 2.31 for a two-dimensional representation of the problem). Unfortunately the geometry of AFM probes is not well known from first principles, as they do not normally have simple geometrical shapes. Furthermore, because the method of manufacturing such probes is not perfectly reproducible, there will be variability from batch to batch and even considerable variation in probe characteristics produced on the same silicon wafer. Thus, in order to correctly deconvolve the tip contributions from an image, it is necessary to characterize each tip on a case by case basis.

Unfortunately, it is difficult to find a technique to accurately and non-destructively measure the tip geometry. Both TEM and SEM have been tried [45–47], but although either technique could have in principle sufficient resolution for the task, there are a number of difficulties. Neither technique gives real three-dimensional information, so superposition of various rotated images is necessary to fully characterize the probe. The resolution required must be similar to that of the details in the image, i.e. the same as the resolution of AFM. To achieve such resolutions, electron microscopes must be operated at high driving voltages, and with such high voltages sharp features such as the probe are easily damaged by over-charging, especially without metal coating, which can alter the profile of the tip. Such difficulties mean that these techniques are quite impractical. The ideal instrument to get three-dimensional, high-resolution measurements of the probe is the AFM itself. This can be used by tip self-imaging which is done by using the AFM to image a spike-like feature (see for example the tip-characterization samples described in Appendix A). This results in the spike imaging the tip, i.e. the resulting image is an image of the tip, only somewhat dilated by the sample, assuming that the radius of the spike is much less than that of the tip. A related technique is to image a very well-known sample such as well-characterized nanoparticles [48, 49]. Knowing the correct sample geometry the dilation by the tip is easy to extract. These techniques are somewhat more convenient

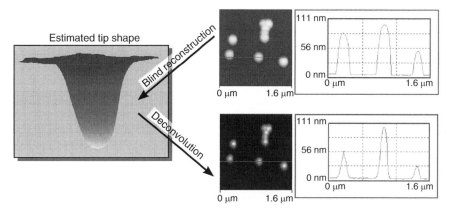

Fig. 2.32. Example of blind reconstruction and subsequent deconvolution of probe geometry. By using the blind reconstruction technique, it is possible to calculate the probe geometry from the image of the nanoparticles (top). Then the probe geometry can be removed from the entire image, giving the sharper image shown below.

than electron microscopy, but both involve imaging an external sample, and could alter the nature of the tip by contamination (e.g. by the nanospheres) or wearing (by the spike), and the stability of a very sharp feature to repeated imaging by the AFM is uncertain [50].

However, using a technique called blind reconstruction, the geometry of a probe can be calculated by carefully analysing an image of the sample of interest [50–53]. Blind reconstruction works by assuming that the AFM image is formed by a dilation process [54]. This means that not only does the tip probe the sample, but the sample also probes the tip. In principle, each pixel in the image obtained contains information about the shape of the AFM probe tip, as well as the information about the real sample topography. The idea of blind reconstruction is to extract the information about the tip from each pixel. In order to build up a picture of the tip from the information in the image, the correlation between the neighbourhood of each pixel is compared to the neighbourhood of each other pixel in the image. Essentially, the routine looks for any repeating patterns in the image, and assumes that any such patterns derive from the shape of the tip. Such methods are very computer intensive and it can take a long time to analyse a single image. For example, to do the blind reconstruction on a single 512×512 pixel image with a Core2 processor can take 5 minutes. Having obtained the geometry of the tip from the image, it is then possible to mathematically remove the probe broadening artefacts from the image. As an example, Figure 2.32 shows an AFM image of 100 nm diameter nanospheres, and the effects of blind tip-shape reconstruction and subsequent image deconvolution on the image.

This technique has been shown to be highly effective under certain circumstances [45, 55], but it does have some limitations [54]. The tip profile extracted is not the real profile, but an estimate of the profile of the bluntest tip that could have made the image. In addition, imaging some samples will give a tip profile estimate that is closer to the real shape than others. Specifically, samples with high roughness and steep features will lead to a tip profile close to the real one. If the sample measured does not have these, it may be necessary to image an external sample after imaging the sample of interest to get adequate data for blind reconstruction. Also, the accuracy of the estimate was shown to depend strongly on the image size used. Specifically, small images will tend to overestimate the sharpness of the tip, due to lower signal-to-noise ratio at small scan sizes and this noise at high resolutions can also introduce distortions in the estimated tip shape versus the real shape [56]. On the other hand, large images or images with a low pixel density will tend to give results suggesting a much 'blunter' profile of the tip due to undersampling of the surface topography [57]. Overall, blind reconstruction is a technique which is relatively simple to apply, and can considerably improve the quality of AFM data, by removing large part of the dilation effect of the tip, as long as care is taken in the application of the technique. It is always worth remembering, however, that in the case of a blunt tip or sharp sample features, there will exist some parts of the sample that are never probed by the tip [58] (see for example, Figure 2.28, where the corners of the depressions, or corners of the protrusions could not be imaged), and deconvolution cannot help here.

2.5.5   Cantilever force constants

Cantilevered AFM probes fabricated with MEMS processing technologies are subject to fairly dramatic distributions of specifications. For example, a wafer of probes could have only 85% of the probes less than 10 nm in diameter. The other 15% could have any diameter. This distribution can dramatically affect the quality and resolution of AFM images. Besides the probe geometry, the cantilevers can have a substantial variation in specifications, especially the thickness of the cantilevers. Because the force constant of the cantilever varies as the inverse cube of the thickness, the force constants of MEMS fabricated cantilevers can vary dramatically. Table 2.1 shows an example of the variability of cantilever geometries and the calculated impact on critical specifications. It can be seen that in relative terms, the thickness has a very high variability compared to the width or length, and that this greatly affects the force constant variability.

Calibration of cantilever force constants
As illustrated in Table 2.1, there is considerable variability in the force constant of an AFM cantilever, mainly because of variations in the thickness of the cantilevers. Thus, when knowledge of the exact force applied is required, the force constant of the cantilever used for the tests must be calculated [59]. This is particularly important for force spectroscopy applications (discussed further in Chapter 3), but knowledge of the applied force is also very important for reproducibility in imaging studies. A large amount of work has been done to determine optimum techniques to measure the force constants of AFM

Table 2.1. Example of the variability in manufactured AFM cantilever properties.

Cantilever Data	Value	Range
Thickness	2 μm	1.5–2.5
Mean width	50 μm	45–55
Length	450 μm	445–455
Force constant	0.2 N/m	0.07–0.4
Resonance frequency	13 kHz	9–17

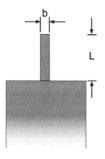
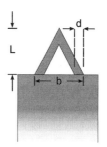

$$K_{rec} = 0.1906 \rho b^2 \, LQ_r \omega_f^2 \, \Gamma_i^f \, (\omega_f) \qquad (2.11)$$

$$K_{tri} = \frac{D_0 \overline{d}}{2L^3} \left(1 + 4 \frac{d^{-3}}{b^3} \right)^{-1} \qquad (2.12)$$

where:

k_{rec} = rectangular cantilever force constant

k_{tri} = trianguar cantilever force constant

ρ = density of the medium

Q_r = Quality factor at resonance

ω_f = Resonant frequency

$\Gamma_i^f \, (\omega_f) =$ imaginary part of hydrodynamic function

$$D_0 = k_{rec} \frac{4L^3}{b}$$

$$\overline{d} = \frac{d}{\sqrt{1 + \frac{b^2}{4L^2}}}$$

Fig. 2.33. Equations for calculating force constants of rectangular and triangular cantilevers with the Sader method [66].

cantilevers (see, for example references [60–65]). The most commonly used method used for to calculate the normal force constant is the known as the Sader method [66–68]. In this method, the physical geometry of the cantilever is measured, typically with an optical microscope and the quality factor, Q, is measured. The quality factor is of the cantilever is essentially a measure of the bandwidth of the oscillation compared to the exciting frequency), and is simple to measure with the AFM. With the physical measurements, and the Q of the cantilever, Equations 2.11 and 2.12 can be used to calculate the spring constant, see Figure 2.33. This method has been later extended to the calculation of k_{tor}, the torsional constant [61], as well as to cantilevers of arbitrary geometry [69].

Several other methods for cantilever calibration are also widely used. The added mass method, first described by Cleveland [70], takes advantage of the change in frequency of a lever when a mass is added to the end. Essentially, the method consists of physically adding a small well characterized particle to a probe tip, for example a tungsten or polymer microsphere. This method is at least as accurate as the Sader method [63, 71], and does not require measurement of cantilever dimensions, but although described as non-destructive, is hard to perform without the danger of damaging the tip, and relies on having well-characterized masses readily available.

The third method commonly applied to obtain the normal spring constant is known as the thermal noise method, first presented by Hutter *et al.* [72, 73]. This method requires measuring the thermal noise spectrum when the cantilever is in contact with a surface. The thermal noise spectrum can be very weak, and it is necessary to use a high performance spectrum analyser to collect the data, pressing the cantilever against the surface means it can potentially change the tip shape. However, the technique has been shown to be both precise and accurate [62].

Finally, the reference cantilever method [74, 75] consists of pressing the cantilever against a pre-calibrated spring, or reference cantilever. This is relatively easy to perform, but relies on the availability of a well-characterized cantilever, and again, could be considered to alter the tip shape.

All the methods mentioned above have focussed on the normal force constants of cantilevers. However, for certain applications torsional or lateral force constants must be calibrated [76, 77]. This is discussed further in Chapter 3, in the sections on lateral- and torsional-bending based modes.

2.5.6   Improved probing systems

The scan rate of an AFM limits the sizes of areas that can be analysed to a few hundred microns at best. It would be highly desirable to create an AFM with multiple probes that could scan many areas simultaneously. Several efforts to create 'multiple probe' atomic force microscopes demonstrated that it is possible. In the first approach, several AFM scanners were positioned above a silicon wafer sample and scanned independently. In the second approach, several probes on one silicon wafer were used to scan a sample simultaneously [7]. The greatest challenge for creating multiple probe AFM instrumentation is to get all of the probes to be as sharp as is required for high-resolution scanning. In most multiple probe AFM designs, a simplified cantilever actuation/sensing system is used, to avoid the complication of fitting multiple optical lever set-ups around closely-spaced cantilevers [78]. These set-ups also simplify the operation of the instruments greatly, although they typically mean that standard cost-effective cantilevers cannot be used.

Another approach to improve the overall productivity of AFM is to maintain a single probe, but to drive it at much higher speeds than are available by normal AFM. As shown in Figure 2.29, standard AFM cantilevers have lengths of the order of several hundred micrometers, typically 200–400 μm, and widths of the order of 40 μm. However, there are advantages to be gained in specific applications in using much shorter and smaller cantilevers [3, 79–81]. Fast scanning AFM benefits from the use of small cantilevers, specifically because they can have much higher resonant frequencies than larger cantilevers, without having very high spring constants, and can cause less sample damage when scanning at higher speeds [3, 34, 81]. Small cantilevers also have advantages for force spectroscopy, because the noise floor is typically controlled by perturbations of the cantilever by the surrounding medium (air or liquid), which has a smaller effect for smaller cantilevers, leading to higher signal-to-noise ratio [82]. Small cantilevers for fast scanning AFM were first proposed in 1996 [83], but their use is far from widespread. The main problem is that use of such cantilevers in a normal optical lever AFM head needs modifications to the optics of the light lever [3, 84]. Nevertheless, the use of small levers for both spectroscopic and fast scanning applications is growing, and recently a commercial source of such levers became available, and commercial AFM instruments capable of using laser spots sizes suitable to use small levers are now also available.

2.6   AFM instrument environment

Various environmental factors can drastically affect the results obtained with AFM. These include the environment the instrument is maintained in, for example the presence of vibrations in the building in which the AFM is housed. An AFM is a very sensitive device, and is highly susceptible to interference from vibration. The two main types of vibration picked up by the AFM are acoustic and mechanical vibrations. Acoustic vibrations are transmitted through the air, and are sometimes, but not always, also audible to humans. In deciding where to place an AFM in your organization, it is always worth looking for the quietest room you can find; even people talking quietly can show up in the AFM when scanning at very high resolution, so it is ideally located in a place with few other activities taking place. Section 6.4.2 shows an example of acoustic noise in an AFM image. In general, locating the AFM in a room with little traffic and audible noise reduces these sorts of interfering vibrations to a reasonable level. Some acoustic noise however, such as the fans used to cool the AFM or computer system, or from air conditioning units are hard to avoid. In addition, increasingly, AFM instruments are being installed in locations with unfavourable characteristics for the operation of an AFM. In these cases, some form of isolation from acoustic and floor vibrations is necessary. Typically, the AFM is easily isolated from acoustic noise by placing it in a sealed cabinet constructed from acoustic damping materials. One drawback of these cabinets is that such materials are commonly made of polymer foam, and they tend to be good thermal insulators as well, so that a small acoustic cabinet can often mean it is more difficult to disperse the heat generated by the AFM, leading to poor thermal stability. An alternative to such cabinets is to clad the walls of the room with sound-absorbing panels; these greatly decrease the transmission of acoustic noise to the instrument without reducing the cooling potential.

Mechanical vibrations are transmitted to the AFM stage from mechanical contact. There are many possible sources of such mechanical noise. Typically, however, the noise reaches the AFM via floor contact. Therefore isolation of the AFM stage from the floor is necessary. The most simple method to do this is by simply mounting the AFM on a granite stage (often included as part of the AFM), and further isolating with a metallic plate/rubber balls stack. Such systems work adequately for low-resolution work or work with quite rough samples, but are generally inadequate for more demanding applications. For better vibration isolations two types of solution exist; active and passive vibration isolation.

Active vibration solutions generally take the form of a platform for the AFM, which may be suspended, devices that measure the vibration transmitted from the base of the platform (accelerometers), and active elements (transducers) that counter measured vibrations, via a feedback loop including the signal from the accelerometers. Such devices work extremely well in isolating AFM instruments, and are generally very compact.

Fig. 2.34. Examples of different types of vibration isolation. Left: an active vibration isolation platform (reproduced with permission from Herzan). Centre: a compressed air-based isolation table. This type of passive vibration isolation is particularly appropriate for large instruments. Reproduced with permission from Kinetic Systems, Inc. Right: an image of an AFM mounted on a bungee cord suspended platform. The suspended platform is enclosed in a cabinet designed to further shield the AFM from acoustic noise. Image used with permission from Agilent. Metallic springs are also suitable for suspension of platforms.

Passive vibration solutions basically consist of some sort of spring, a weight, and a damping device. Usually, these will take the form of an air table or a suspended platform. The suspended platform is a very simple device, and consists of placing the AFM on a heavy stage (which forms the weight) that is suspended by springs (either strong elastic bungee cords, or metal springs). With long springs and a large weight, these devices have excellent isolation characteristics, but use a lot of space. The air table is marginally more compact and consists of a suspended platform, which is cushioned by air. In general, any of the solutions mentioned here work extremely well, and attenuate floor vibrations from most locations to an acceptable level for high-resolution AFM work. Some typical devices are shown in Figure 2.34.

2.7 Scanning environment

Typically AFM is carried out under ambient conditions, but a major advantage of AFM is the ability to image the sample in almost any environment, from vacuum, to gas, to liquid. The most common application is to scan in liquid, which is particularly important in biological applications in order to measure the samples in their native state, or for electrochemical measurements of metallic surfaces. For more examples of liquid scanning applications see Chapter 7. In addition, scanning in liquid has some advantages over scanning in air, such as reproducibility of the scanning environment, and notably, the tip–sample interactions in liquid are much weaker. One of the major problems of scanning in ambient air is that typically a meniscus layer of water will coat the sample and/or tip. Interaction of the AFM with this layer of water will dominate the behaviour of the tip–sample interaction [85], making it difficult to apply a small force to the sample when scanning (especially in contact mode) [86]. Scanning in water removes these forces altogether.

Practically, most AFMs require the use of a specific liquid cell in order to scan in fluid. The leaking of water in an AFM experiment can cause problems for the instrument. Apart from the normal incompatibilities between electronics and water, piezo-electric scanners are highly sensitive to moisture and must be well sealed to prevent

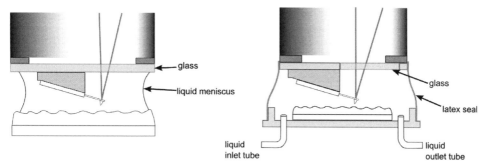

Fig. 2.35. Two approaches to AFM scanning in liquid with a probe-scanning microscope. Scanning in a droplet of water (left) is simpler, but allows less control of the solution. Scanning in an AFM fluid cell (right) allows exchange of fluids, and control of fluid temperature, ionic strength, etc.

damage from water from the liquid cell. Due to the specific danger of piezo damage, AFMs designed for liquid work are usually based on the tip-scanning design, to minimize the chance of liquid reaching the scanner. AFM liquid cells will typically have ports to exchange fluids during scanning, and can be of a semi-closed or completely sealed design. With sealed cells, some flexibility (a rubber membrane or o-ring) must be included to permit the movement of the piezo, or of the sample. Other typical additions to liquid cells are electrical connections (for electrochemical AFM), or heating/cooling circuits (see next section). An alternative to a liquid cell for scanning in fluid is to simply scan within a droplet of water. An illustration of this technique and of a closed liquid cell is given in Figure 2.35. In this method, typically applied with probe scanning microscopes, the bottom of the scanner assembly is covered by a glass window, to allow the laser to focus on the cantilever, which is mounted below the glass window. A droplet of liquid sits on the sample surface, and when the scanning head is lowered into it, the liquid forms a continuous bridge between the glass of the scanning head, and the sample surface. This technique only works with aqueous samples, and hydrophilic samples, as the surface tension of the water is required to keep the droplet in place as the AFM head is lowered. Furthermore, the droplet is prone to evaporate over a period of time. With these caveats, however, the technique is surprisingly effective, and greatly reduces the chances of bubbles entering the optical path, which can cause problems in closed cells. Non-aqueous solvents must be scanned in the closed liquid cell. More details about procedures for scanning in liquids are given in sections Chapter 4.

Temperature control
Temperature control of scanning is highly desirable for a wide range of applications in AFM. In fact, AFM has been performed at a very wide range of temperatures; imaging at temperatures as low as 5 K [1, 87] and as high as 750 °C [88]. However, for practical purposes, with commercial AFMs, in air, a more limited range of temperatures is available, due to two main effects:

- at low temperatures – condensation on the sample or microscope;
- at high temperatures – destruction of the AFM scanner.

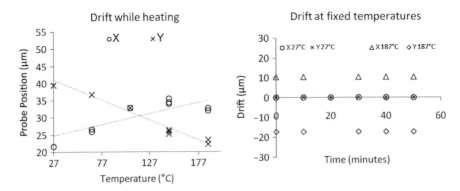

Fig. 2.36. Example data showing the effect of temperature variations on positioning accuracy. Left: while heating, the position of the probe on the sample will drift considerably: Right: at elevated temperatures, once stabilized, the probe will not drift.

When heating the sample, to avoid damage to the scanner the most important thing is to thermally isolate the scanner from the hot sample, this is typically done by use of a spacing material with very high thermal insulation, and high stiffness (i.e. a ceramic such as MACOR® – a ceramic with very low thermal conductivity and a low thermal expansion coefficient).

As a sample is heated or cooled, there will be considerable thermal drift in the AFM stage in the X-Y and Z directions. The thermal drift can only be minimized by making the stage as symmetrical as possible and using low drift materials. Usually there is drift as the stage temperature is raised, and the drift stops when the desired temperature is reached and the stage temperature is stabilized. Data illustrating this is shown in Figure 2.36.

Cooling samples in an AFM stage has another potential problem. Condensation occurs on the components that are being cooled. When cooling a sample in an AFM stage, it is recommended that the entire stage be maintained in a dry environment such as in a glove box. If going to very low temperatures, the ideal situation is to scan in a vacuum. As such temperatures are generally used in physics applications this is rather common. Another approach is to image in liquid nitrogen vapour at ambient pressure [89]. At cryogenic temperatures, AFM becomes very sensitive, due to the lack of thermal excitation from air molecules on the cantilever, and also reduced sample movement, and some unique experiments may be carried out. Examples include biological molecules too delicate to measure in ambient conditions [90, 91], measurement of frozen liquids [92] and even single atom hopping from sample to tip [1]. While the design and use of AFMs to be inserted in vacuum and operated at cryogenic temperatures is outside the scope of this book, the interested reader is directed to the references [90, 93].

Chapter 3

AFM modes

The range of available AFM modes and experiments are at the heart of modern AFM. The wide variety of experiments that may be performed with an AFM make it a versatile, powerful tool. Initially the only mode available for imaging was contact mode, and this limited the types of samples that could be examined, the types of experiments that could be performed, and the type of data that could be produced. Now there are a very large number of possible modes of operation of AFM. For example, in 1999, Friedbacher *et al.* attempted to list the names of all the SPM modes described, and arrived at more than 50 terms [94]. Some of these SPM modes were STM or SNOM (Scanning Nearfield Optical Microscopy), but there are *at least* 20 different modes of AFM. SNOM is an example of using the close contact and position control of the AFM to measure properties of the surface other than topography (in this case, optical properties). As shown in this chapter, many of newer modes in AFM are along these lines: techniques that use the incredible resolution provided by scanning a probe close to the surface with an AFM to measure different properties of the sample surface on the nanoscale. SNOM (also sometimes called NSOM, near-field scanning optical microscopy), is a very powerful technique combining near-AFM resolution with the spectroscopic information that is available by using light-based techniques. However, SNOM is not covered in this book because although some early SNOMs instruments were developed by modification of AFMs, experiments are now generally carried out with specialized SNOM instruments, which are rather different from a normal AFM. For reviews of SNOM, see [95, 96]. A table categorizing major techniques in Scanning Probe Microscopy is shown in Figure 3.1.

For the purposes of making this a practical guide to AFM, in this chapter we concentrate on techniques likely to be accessible and of interest to the reader. This means that we will not describe in detail techniques which are not attainable with commercial AFMs without significant modification, nor cover modes that have been described but not widely adopted. Some more advanced techniques are covered in the applications section, Chapter 7.

3.1 Topographic modes

The basis of AFM as a microscopic technique is that it measures the topography of the sample. As described in Chapter 1, the datasets generated in this way are not conventional images, as produced by optical microscopy, but rather a map of height measurements. These may be later transformed into a more naturalistic image with light shading, perspective, etc. to help us picture the shape of the sample (this process is covered in Chapter 5). In order to make these height measurements, a variety of modes have developed, which can be divided into those modes which measure the static deflection

AFM	SNOM
Contact mode	Aperture (ASNOM)
Non-contact mode (NC-AFM, close contact mode, FM-AFM)	Non-aperture SNOM (NA-SNOM)
	Evanescent field SNOM (EF-SNOM)
Intermittent Contact mode (IC-AFM, AM-AFM, Tapping)	Transmission SNOM (T-SNOM)
	Collection SNOM (C-SNOM)
Chemical Force Microscopy (CFM)	**STM**
Lateral Force (LFM, FFM))	Scanning Tunnelling Spectroscopy (STS)
Electric Force (EFM)	Topography (STM)
Force Spectroscopy	Alternating Current STM (AC-STM)
Nanoindentation	Ballistic electron emission microscopy (BEEM)
Magnetic Force (MFM)	Scanning Tunnelling Optical Microscopy (STOM)
Kelvin Probe (KPM, SKPM)	
Scanning Thermal Microscopy (SThM)	
Nano oxidation Lithography	
Dip-pen Nanolithography (DPN)	

Fig. 3.1. Summary of the names of some SPM-based techniques.

of the AFM cantilever, and those that measure the dynamic oscillation of the cantilever. The differences between the modes lead not only to different experimental procedures, but to differences in the information available, differing suitabilities for particular samples, and even to differences in the interpretation of the data.

3.1.1   Contact mode

Contact mode AFM was the first mode developed for AFM. It is the simplest mode conceptually, and was the basis for the development of the later modes. Therefore understanding contact mode helps to understand how the other techniques work. Although the limitations of contact mode prompted the development of modes that could examine different samples in different environments and give different information, contact mode is still an extremely powerful and useful technique. Contact mode is capable of obtaining very high-resolution images. It is also the fastest of all the topographic modes, as the deflection of the cantilever leads directly to the topography of the sample, so no summing of oscillation measurements is required which can slow imaging.

In order to understand the way AFM modes work, it is necessary to use so-called force–distance curves. A cartoon of a simple force–distance curve is shown in Figure 3.2. As the name implies, these curves are a plot of force (on the y axis) versus distance (on the x axis). Such a curve is simple to acquire with the AFM. It is calculated from a deflection–distance curve which is easily measured by monitoring the deflection of the cantilever as the piezo is used to move the tip towards the sample. Typically, at a set deflection level, the direction is reversed, and the tip withdraws from the sample. This results in a deflection versus distance curve, which may be converted to a force–distance curve. Measurement of force–distance curves is a very sensitive and quantifiable way to

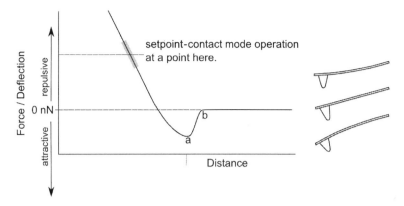

Fig. 3.2. Simplified force–distance curve showing contact (repulsive region) scanning regime. A deflection–distance curve, which is the raw data from which a force–distance curve is measured, has a similar shape. Right: illustration of probe bending in each regime.

determine tip–sample interactions, and is the basis for several non-topographic AFM modes, such as force spectroscopy and nanoindentation. More information about the use of such capabilities of force–distance curves, and how they are converted from deflection to force is given in Section 3.2.1, and Section 4.5, respectively.

Considering the approach curve shown in Figure 3.2, when the tip is far from the sample surface, the cantilever is considered to have zero deflection; as the tip approaches the surface, it normally feels first an attractive force, and a 'snap-in' occurs, as the tip becomes unstable and jumps into contact with the surface. As the instrument continues to push the cantilever towards the surface, the interaction moves into the 'repulsive' regime, i.e. the tip is now applying a force to the sample, and the sample applies an opposite force to tip. In this regime, a combination of cantilever bending and sample compression will be occurring according to the relative compliances of the sample surface and AFM probe. If the direction of movement is reversed, the interaction passes again into the attractive regime, and the tip stays on the surface until instability occurs once more, and the tip snaps off the surface. It is within the repulsive regime that contact-mode imaging usually occurs (for example, at the point labelled 'set-point' in Figure 3.2). In other words, in contact-mode AFM, the tip of the probe is always touching the sample. This has the following important implications for contact-mode AFM:

1. As a result of the repulsive force between the tip and the sample, the sample may be damaged or otherwise changed by the scanning process.
2. Conversely, the tip could also be damaged or changed by the scanning process.
3. As the tip and sample are constantly in contact with each other as the tip moves along the sample, in addition to the normal force they apply to each other, lateral forces are experienced by both probe and sample.
4. The contact between the tip and the sample means that the nature of the sample surface may affect the results obtained. This means that the technique can be sensitive to the nature of the sample.

The forces applied to the surface by the probe in contact mode are given by Hooke's law:

$$F = -k \times D \tag{3.1}$$

where F = force (N), k = probe force constant (N/m) and D = deflection distance (m).

The basis of contact-mode AFM is that the microscope feedback system acts to keep the cantilever deflection at a certain value determined by the instrument operator. This point is known as the set-point. The set-point is one of the important control parameters that the operator must adjust to optimize imaging, and there are equivalent parameters for all other AFM imaging modes as well. Equation 3.1 shows that either a probe with a high force constant (one with a stiff cantilever), or a greater deflection (i.e. a higher set-point), will lead to a higher applied force. Because the feedback system of the AFM cannot have instantaneous response, the vertical deflection will actually vary somewhat during imaging (indicated by the red region of the curve in Figure 3.2). The amount it varies will depend on the topography of the sample, flexibility of the cantilever, scanning speed, and how well the feedback circuit has been optimized. Optimization of these parameters is discussed in Section 4.2. The AFM software may display the deflection signal as a line plot as the tip passes over the sample, or as an image. The deflection signal in contact-mode AFM is the error signal, that is, the size of the deflection is a measure of how much the cantilever is deflecting *before* the deflection is 'corrected' by the feedback circuit via height adjustment by the piezo. Therefore, in the ideal situation, there would be *no* contrast in the deflection image. The more contrast exists in the deflection signal, the more 'errors' will be present in the height image, because regions of high contrast in the deflection image correspond to regions in the height image, where the feedback has not yet corrected for cantilever deflection. However, usually it is not possible to have the feedback signal respond perfectly, and the deflection signal will show the slope of the sample, because it is regions where there is high slope, or more precisely, a great rate of change of slope with distance, that give rise to large cantilever deflections.

The imaging mode described so far is known as *constant-force* contact-mode AFM. If the user turns off feedback altogether while imaging, then they are effectively using *constant-height* contact-mode AFM rather than *constant-force* contact-mode AFM. Because constant-force mode is by far the most widely used mode, in general any reference to contact-mode AFM will mean constant-force mode unless specified otherwise, and this is the convention we follow in this book. In constant-height mode, with no feedback active, the image signal comes entirely from cantilever deflection, rather than from the voltage applied to the *z piezo* (which would be typically set at a constant value). Height measurements will therefore require specific calibration of the cantilever deflection, to extract real sample topography. Constant-height mode AFM does have some specific applications: it can be useful in conditions where scanning is carried out so fast that the feedback system cannot cope [9, 36]. However, under these conditions, AFM is in fact acting rather like a stylus profiler as the tip–sample force is not fully controlled. Typically, to carry out these measurements, the feedback system is initially used to determine the location of the sample surface, and the approximate topography, before being turned off, or just reduced to a very low level.

However, even using constant-force AFM, the software typically will allow the user to save the deflection image, and some operators choose to publish this image, as it is often a

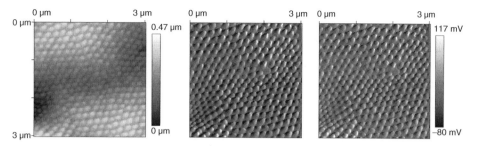

Fig. 3.3. Illustration of the relation between height and deflection images. Left to right: height, right-shaded height and deflection images of the surface of a mosquito eye. The height image shows how much the z scanner moves to maintain the set-point. The deflection image shows how the cantilever bends as it passes over the sample, and is the signal used for feedback in contact-mode AFM. This image is very similar to the shaded height.

simple way to show the shape of the sample, and may even show features not visible in the height image (which could be an indication that feedback was not optimized). However, it is worth remembering that where feedback was correctly optimized, the AFM height image will also contain all the features present in the error signal. One way to show this is to apply a shading algorithm to the height image – this effectively gives the derivative of the height image; the resulting image will be very similar to the deflection image. An example of this is shown in Figure 3.3. Note that in the deflection image shown in Figure 3.3 the z-scale is in volts. The z-scale was included here for illustrative purposes, in such images the z-scale is almost completely meaningless scientifically – even if converted to nanometres, the size of the deflection could easily be changed by adjustment of the feedback parameters, and so should always be removed from the image before presentation.

The deflection signal is used, as described previously, with the feedback parameters to determine how the z piezoelectric must move to maintain a constant cantilever deflection (and hence constant tip–sample force). The amount the z piezo moves to maintain the deflection set-point is taken to be the sample topography; this signal, plotted versus distance, forms the height or topography image in contact-mode AFM. There is a third signal which is typically available in contact-mode AFM. This derives from the lateral twisting of the cantilever, and is therefore usually called lateral deflection. This signal is typically used due to its material sensitivity, rather than as a measure of sample topography, and it is therefore covered in the non-topographic modes part of this chapter (Section 3.2.3.1). The origin of the vertical and lateral deflection signals in a typical optical lever AFM set-up is shown in Figure 3.4. The photodetector used in optical lever AFMs usually comprises of four segments. The difference in signal between the top two and bottom two segments, i.e. $(A + B)–(C – D)$ gives the vertical deflection (measured in volts, or amps), and the difference between the rightmost two segments and the leftmost two segments gives the lateral deflection, i.e. $(B + D)–(A + C)$.

It is worth noting here that in contact-mode AFM, like most of the other modes, two versions of each data type can be available, these being the data collected in the left-to-right direction, and the reverse set, collected in the right-to-left data direction. By

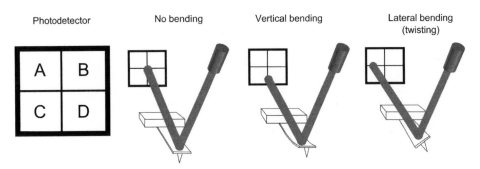

Fig. 3.4. Illustration of how the photodetector detects vertical and horizontal bending of the cantilever.

convention AFM images are presented with the fast scan axis data appearing horizontally, no matter in which direction the AFM tip was scanned. So there will be forward and backward height, vertical deflection and lateral deflection images available, meaning up to six data channels might be recorded. If the instrument is equipped with a z axis calibration sensor, both z voltage and z sensor channels might be available, raising the total to eight channels. Typically, while the probe scans over each line in both directions, only one direction will be saved. This is because the height data in the two directions should be identical. The vertical deflection images should be the same on parts of equal slope, and give opposite contrast on regions of changing slope, but the information available from the data collected in the two directions is effectively the same. So for most channels, there is rarely any need to save the data collected in both directions, although it is sometimes useful to observe both forward and backward height data while optimizing scanning conditions, as discussed in Section 4.2. Lateral deflection data from both directions is sometimes saved, to help understand frictional properties of the sample, which is discussed in Section 3.2.3.1.

Applicability
Contact-mode AFM has a really wide range of potential applications and some of these are described in Chapter 7. However, it is possible to summarize some general cases where contact mode is likely to be chosen in preference to other techniques. Probably the best reason to use contact mode is its high resolution. Some dynamic modes can also achieve extremely high resolution, but compared to, for example, intermittent-contact mode, the resolution of contact mode is potentially extremely high. What keeps it from being used more widely is that the applied normal force leads to a high lateral force applied to the sample as well. In the case of weakly adsorbed samples, or soft, easily deformed samples, this can lead to problems of sample distortion, damage, or even removal from the substrate [97, 98]. Because of this, it has been suggested that contact-mode AFM is no good for soft samples. This is not the case, as there are many reports of soft biological samples, even in a hydrated state, being successfully imaged by contact-mode AFM (for examples, see [99–103]). Often, contact mode is chosen to image such delicate structures when sub-molecular resolution is required [99, 100, 102, 104]. What

is true is that in ambient conditions, a capillary layer of water will form between the tip and the surface. One effect of this is to 'pull' the AFM tip onto the surface, often applying an even stronger force than the force applied (via the set-point) by the operator. Thus it is easy to unwittingly apply a very large force (>nN) to the sample in contact-mode AFM in ambient conditions. In water, these forces do not exist, so it is easier to image with a very gentle force. For this reason, and due to some complications of imaging in dynamic modes in liquids (see the next section), imaging in liquid is a strong point of contact mode. As mentioned previously, contact mode also works well in high-speed AFM, and some high-speed AFM set-ups use this mode exclusively [36].

3.1.2   Oscillating modes

In the first paper on AFM, Binnig and Quate acknowledged the potential benefits of oscillating the cantilever in an AFM, and compared the results of using an oscillating probe with those from contact mode. At the time contact mode gave far better results, probably due to the nature of the probe used [19]. Although the use of oscillating modes were revisited shortly afterwards [105], it was several years before oscillating probe modes became popular, and for quite a while nearly all AFM was carried out in contact mode. The primary motivation for using oscillating mode in an AFM is to take advantage of the signal-to-noise benefits associated with modulated signals. Thus, an AFM that has oscillating modes can measure images with a small probe–sample force.

There are now a large number of dynamic modes of operation, and even more names for those modes. However, all of these modes are variations on a theme. The cantilever is oscillated, usually with an additional piezoelectric element, and typically at its resonant frequency. When the oscillating probe approaches the sample surface, the oscillation changes due to the interaction between the probe and the force field from the sample. The effect is a damping of the cantilever oscillation, which leads to a reduction in the frequency and amplitude of the oscillation. The oscillation is monitored by the force transducer (i.e. by the optical lever in most AFMs), and the scanner adjusts the z height via the feedback loop to maintain the probe at a fixed distance from the sample, just as in contact-mode AFM. The only real differences between the various oscillating modes available are in the size (amplitude) of the oscillation applied to the probe, and the method used to detect the change in oscillation. The general principle of oscillating AFM modes is shown in Figure 3.5.

Irrespective of the many different terms used to describe the techniques, there are actually only a few kinds of conditions used in oscillating imaging modes. The user can decide to set either a small or a large applied oscillation amplitude, and sometimes can decide how to detect the change in probe oscillation. Some instruments may only have one detection scheme implemented. The instrumental set-up schematic is shown in Figure 3.5. An oscillating signal is generated, and applied to the cantilever mechanically, such that the probe is oscillated close to its resonant frequency. The oscillation of the probe is monitored as it is brought close to the sample surface. The detected change in the oscillation (whether detected via amplitude, phase or frequency), is used in a feedback loop to maintain the probe–sample interaction constant. The choice of small or large amplitude has a considerable practical effect, as is illustrated in Figure 3.6. Using a small oscillation amplitude (Denoted by the arrow **A**), it is possible to maintain

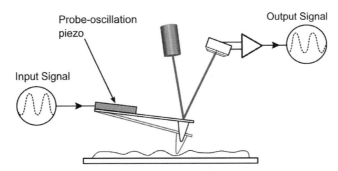

Fig. 3.5. Schematic of generalized operation of oscillating AFM modes, showing instrumental set-up. An oscillating input signal is applied to the cantilever to make the probe vibrate up and down. The actual movement of the probe will depend on its interaction with the sample surface. The resulting oscillation in the cantilever deflection is measured and compared to the input oscillation to determine the forces acting on the probe.

the cantilever in the attractive regime only. This technique is sometimes known as non-contact AFM, or alternatively, and perhaps more accurately, as close-contact AFM (see Table 3.1). This technique has some advantages due to the low probe tip–sample forces involved, and is discussed below in Section 3.1.2.1. On the other hand, it can be seen that if a large oscillation amplitude is applied, then the probe will move from being far from the surface where there's no tip–sample interaction, through the attractive regime, into the repulsive regime, and back, in each oscillation cycle (arrow **B**). This technique involves large probe tip–sample forces, so can be more destructive, but is easier to implement. This technique is what we call intermittent contact-mode AFM (and is also known by many other names, some of which are given in Table 3.1), and is discussed in Section 3.1.2.2.

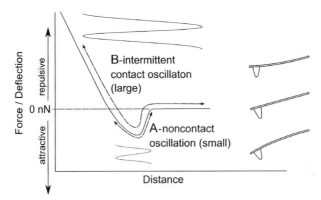

Fig. 3.6. Different operating regimes for oscillating AFM modes. A: with a small amplitude of oscillation, the probe can be kept in the attractive regime. B: with a larger oscillation the probe moves through non-interacting, attractive and repulsive regimes, resulting in intermittent contact.

Table 3.1. Nomenclature of some oscillating probe AFM modes.

Detection	Amplitude	
	Low	High
Amplitude	Rarely used	Intermittent Contact AFM (IC-AFM), also known as AC-AFM or Tapping
Phase	Non-contact AFM (NC-AFM), also known as close-contact AFM	Rarely used

Typically an AFM designed for use in air or liquid has electronics that can measure changes in vibrational amplitude or phase at a preselected frequency. So the instrument operator can choose to use either of these for feedback. In combination with large or small amplitudes, there are four types of oscillating experiment available to most AFM users, which are shown in Table 3.1.

It should be stressed that the two possible conditions described as 'rarely used' in Table 3.1 are not unusable, just that they are not commonly applied. Phase detection is usually used with small amplitudes (close-contact AFM), due to somewhat higher sensitivity, and amplitude detection is usually used with large amplitudes (intermittent-contact AFM), but these are not the only possible imaging methods. Optimal imaging conditions are sometimes difficult to establish, and it may be necessary to try different amplitudes and detection schemes to find the ideal conditions.

An alternative to amplitude or phase detection is frequency-modulation detection (FM-AFM), typically used in ultra-high vacuum conditions (UHV-AFM). FM-AFM is typically applied with small oscillation amplitudes in the non-contact regime. Typically FM-AFM is carried out with a phase-locked loop device. This technique is unavailable to most AFM users due to the need for additional equipment, so it is not covered in detail in this book. However, it has been described in detail [106, 107], and compared with the amplitude modulation (AM-AFM) techniques we discuss here elsewhere [108].

3.1.2.1 Non-contact mode/close-contact mode

One of the great advantages of oscillating modes in AFM is that they can decrease the size of tip–sample forces, while maintaining high sensitivity to the sample topography. To achieve non-contact AFM, the tip must be close enough to the sample surface to achieve this high sensitivity, without passing into the repulsive regime used for contact-mode AFM. Non-contact AFM is therefore carried out in the attractive regime, as shown in Figure 3.7.

By using a highly stiff cantilever and monitoring the dynamic effects of the attractive force (i.e. the change in the oscillation) in this regime, it is possible to maintain the cantilever very close to the surface without jumping to the repulsive regime. It is possible to observe changes in the oscillation amplitude and phase in this regime. These effects are caused by a change in the cantilever resonant frequency which is in turn caused by forces from the surface (normally attractive van der Waals forces) acting on the tip. The resonant frequency far from the surface, ω_0 is given by $\omega_0 = c\sqrt{k}$ where c is a function of the

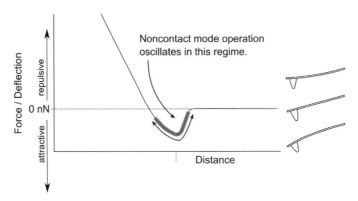

Fig. 3.7. Operating regime for non-contact AFM. With a small amplitude and stiff cantilever, the probe can oscillate within the attractive regime only.

cantilever mass, and k is the spring constant. But an additional force f from the surface means that the new resonant frequency ω_0' is given by:

$$\omega_0' = c\sqrt{k - f'} \tag{3.2}$$

where f' is the derivative of the force normal to the surface [109].

   The important point here is that either the change in amplitude or the change in phase (which actually derives from the change in frequency) may be used in the feedback circuit to maintain the tip at a fixed distance from the sample surface. The name non-contact AFM is actually quite misleading. All AFM modes involve the probe moving into the force field of the sample surface, including 'non-contact' AFM. At the sort of distances involved, it is impossible to say at which point contact occurs. Further misunderstanding is caused by the fact that a number of other names have been used for dynamic AFM modes, and there is no clear consensus on the correct terms to use, so there is great scope for confusion. Here we use the term non-contact-mode AFM to mean AFM carried out in the attractive regime, typically using small amplitudes of oscillation. Section 3.1.2.2 deals with dynamic modes that pass into the repulsive regime, which we choose to call intermittent-contact mode.

Non-contact-mode principles of operation
Typically, non-contact mode is carried out in amplitude modulation mode, and the error signal may be either the amplitude or phase of oscillation of the tip. To avoid the possibility of slipping into the repulsive regime which is likely to damage or contaminate the tip [110], a high-frequency cantilever is typically used with ω_0 in the range of 300–400 kHz. In addition, small oscillation amplitudes are used, often of the order of 10 nm [111]. As with all dynamic modes of operation, scanning speed is usually lower than in contact mode, although the high frequencies and small amplitudes mean scanning speed can often be greater than in IC-AFM. When used in UHV conditions, frequency modulation is usually used [108].

Applicability
Non-contact, or close-contact AFM is a very widely applied technique, and can be used for imaging of almost any sample in AFM. It is currently used less often than intermittent

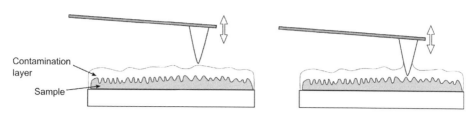

Fig. 3.8. Possible non-contact imaging conditions under ambient conditions, with a sample covered in a contamination layer. Such a layer exists on most samples in air. In the first case on the left, the probe oscillates above the contamination layer. In the second case; right, the probe oscillates within the contamination layer.

contact in ambient conditions. However, with care, it can replace intermittent contact in nearly all applications, and often gives better, and more consistent results due to lower tip wear. One of the limiting factors for non-contact mode in air is the contamination layer present on most surfaces under ambient conditions. In general, the presence of this layer means that the probe–surface interaction forces are governed by the capillary forces between the probe and the contamination layer. For non-contact AFM, The probe may be vibrated in two different distinct regimes as it is scanned across the surface, see Figure 3.8. In the first regime, the probe is oscillated above the surface of the contamination layer. The vibration amplitude must be very small and a very stiff probe must be used. The images of the surface contamination layer are typically unrepresentative of the substrate topography and appear to have low resolution. This is because the contamination fills in the nanostructures at the surface. However, in some cases this technique allows the determination of the location or shape of liquid droplets on the samples' surface, which may be desirable [112, 113]. In the second regime the probe is scanned inside the contamination layer [110]. This technique, sometimes called 'near contact', requires great care to achieve. Again, the cantilever must be stiff so that the tip does not jump to the surface from the capillary forces caused by the contamination layer, and very small vibration amplitudes must be used. However, high-resolution images may be measured in this regime. Non-contact AFM fully immersed in liquid is also possible [114], and delicate samples such as DNA molecules or other biological samples have been imaged by in this way, and such molecules may suffer less distortion when imaged like this than when imaged by intermittent-contact mode [114–116].

Using ultra-high vacuum (UHV) conditions, FM detection has advantages over amplitude or phase detection [117] and FM detection is widely used for UHV non-contact AFM. Some amazing results have been shown for frequency-modulation based non-contact AFM in ultra-high vacuum, including true atomic resolution [118, 119]. For instance, the Morita group have shown true atomic resolution in a number of systems with this technique [106, 118, 120–122]. The system must be very stable for operation to be reliable without the risk of jump-to-contact. An example image showing true atomic resolution by NC-AFM is shown in Figure 3.9. Figure 3.9 also shows a rare example of using NC-AFM to identify atoms on a surface. Force spectroscopy is described further in Section 3.2.1, with respect to using force spectroscopy in contact mode. But in this example, unusual due to the measurement of force curves in FM-AFM mode, force spectroscopy was used to

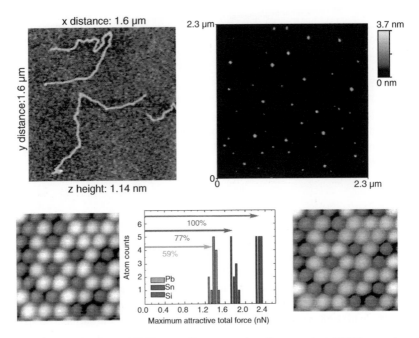

Fig. 3.9. Example non-contact AFM images. Top: examples of non-contact AFM images in ambient conditions (air) – individual DNA molecules (left) and 1 nm nanoparticles (right) [123]. Bottom images: non-contact AFM in UHV conditions for individual atom identification. Left: atomically resolved NC-AFM image of Si, Sn and Pb atoms on an Si(111) substrate – some atoms may be differentiated based on apparent size, but identification is not possible. Middle: short-range chemical force measured over each atom is dependent on the chemical nature of the atoms. Right: the same image as on the left, with atoms coloured according to the colour scheme in the middle. Adapted from [8], with permission. (A colour version of this illustration can be found in the plate section.)

identify the attractive force above individual atoms which could be correlated to their chemical identity [8]. Further examples of the applications of non-contact-AFM to obtain atomically resolved information are given in Section 7.1.5.

3.1.2.2   Intermittent-contact mode

Although the first experiments in dynamic AFM aimed to carry out non-contact AFM, it was not long before the advantages of using a dynamic mode that allows the probe to touch the sample (that is, pass into the repulsive regime) were discovered [97]. For intermittent-contact AFM, feedback is usually based on amplitude modulation [108] and the tip–sample interaction passes from the 'zero-force' regime, through the attractive regime, and into the repulsive regime, as shown in Figure 3.10.

The fact that the tip–sample interaction moves through all three regimes has several important implications:

(i) There is tip–sample repulsive interaction, i.e. tip and sample touch each other, leading to the possibility of sample or tip damage, however:

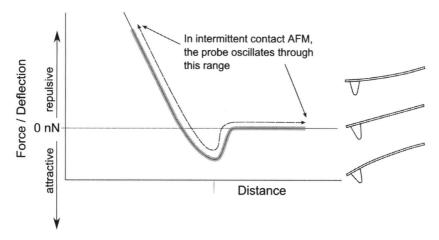

Fig. 3.10. Intermittent-contact operating regime. In this mode, the AFM probe's oscillation is large enough to move from the repulsive regime, through the attractive regime, and completely out of contact in each cycle.

(ii) Due to the movement of the tip perpendicular to the surface as it scans, lateral forces are (almost) eliminated.

(iii) The tip passes through the contamination layer (see Figure 3.11).

(iv) Tip–sample contact also allows some sensing of sample properties.

(v) The feedback system requires the collection of adequate data to characterize the cantilever oscillation in terms of its amplitude.

Points (ii) and (iii) above explain the popularity of IC-AFM. The lateral forces which can cause great problems in contact-mode AFM do not affect IC-AFM. On the other hand, the fundamental instability of non-contact AFM in air (due to operation in the attractive regime, and the presence of the capillary layer) is overcome, making IC-AFM somewhat simpler to achieve. In IC-AFM, the restoring force of the cantilever withdraws the tip from the contamination layer in each cycle, thus reducing the effect of capillary forces on the image.

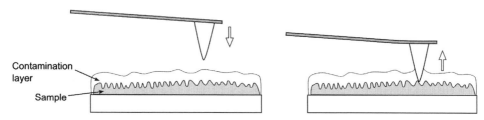

Fig. 3.11. Intermittent-contact-mode imaging conditions in air. The probe passes through the contamination layer to touch the substrate surface, and out again.

Operating principles of intermittent-contact AFM

In IC-AFM the probe is oscillated with a large amplitude, typically in the range of 1–100 nm [108], and the feedback is usually based on the amplitude signal. In most cases, the probe is oscillated by an additional piezoelectric element attached to the probe holder (see Chapter 2), although it is also possible to excite the cantilever vibration by other methods, e.g. by an external magnet, with a magnetically coated cantilever [124, 125], which may reduce fluid vibration when imaging in liquid. In fact, rather than driving the probe directly, the most common excitation method for fluid imaging is to excite the entire fluid cell holder, which causes the liquid to vibrate, acoustically driving the cantilever [126, 127]. Often, in addition to the amplitude signal, the delay in the phase of the probe oscillation is recorded. Oscillation amplitude and phase are illustrated in Figure 3.12.

The amplitude is reduced by the contact with the sample surface, and so an amplitude set-point is set by the user, and the amplitude is the error signal in IC-AFM. In a similar way to deflection in contact mode, the amplitude signal in intermittent contact may be used as an illustration of the shape of the sample. Again, like the deflection signal, the amplitude signal shows where the feedback system has not yet compensated for changes in sample height, so for best height data, the amplitude signal should be minimized. An example image showing the relation between height and amplitude data is shown in Figure 3.13. Note that like deflection images in contact mode, the *z* scale of amplitude images in IC-AFM is usually in volts, unless specifically calibrated. It's common practice to remove this scale for publication as it has no practical use.

In addition to height and amplitude data, the phase-shift may also be saved as an image. The reason why saving this data is useful is not obvious, and this information was largely ignored in early intermittent-contact AFM. In fact, the phase of the oscillating cantilever is strongly affected by the probe tip–sample interactions, so it can be a useful way of distinguishing materials. As a Non-topographic mode, phase imaging is covered in Section 3.2.3.2.

Applicability

Intermittent-contact mode is a very widely applied technique, and is currently the most commonly applied technique for imaging in air. In liquid, IC-AFM mode is also very

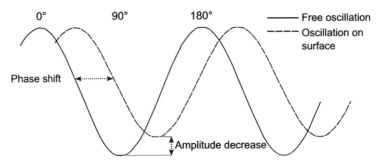

Fig. 3.12. Illustration of the effect of intermittent contact on the cantilevers' oscillation. The free oscillation (solid) is modified when in contact with a surface (dashed) by a reduction in amplitude and a phase shift.

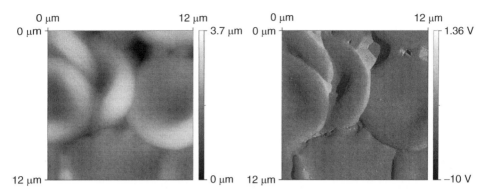

Fig. 3.13. Intermittent-contact AFM images of human red blood cells. Height (left) and amplitude (right) images shown.

widely applied, although it is subject to a number of difficulties specific to operation in liquid, namely that mechanical excitation of the cantilever can lead to excitation of the fluid and fluid cell as well [128], and a lack of clear understanding of the contrast mechanisms [108, 129, 130]. The operation of IC-AFM mode in liquid, as well as in air, is discussed in Section 4.3. Intermittent-contact AFM is not commonly applied in vacuum, due to restrictions in bandwidth due to increase of Q in vacuum [117]. An extremely wide range of samples have been studied by intermittent contact-mode AFM, some of these are illustrated in Chapter 7.

Higher harmonics imaging

A recent development in Intermittent-contact AFM is the use of modes of resonance other than the fundamental one. This may either be by a passive technique, by measuring the vibration at these higher modes, can involve excitation at multiple frequencies. Addition of such capabilities to an AFM is relatively simple, the main requirement being that a lock-in amplifier capable of monitoring the very high frequencies. Figure 3.14 shows illustrations of the first four modes of a beam-shaped cantilever. The requirement for a high-frequency amplifier is because higher modes of real cantilevers are likely to have extremely high frequencies. Because the modes are anharmonic, the second mode is not necessarily at double the frequency of the fundamental (i.e. $f_2 \neq 2f_1$), but may be as high as six times the fundamental frequency [131]. In any case, having two lock-ins is useful because it is advantageous to be able to monitor both f_1 and f_2 simultaneously.

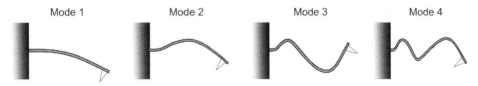

Fig. 3.14. Illustrations of the first four normal resonance modes of a beam-shaped cantilever.

The reason for interest in monitoring the higher modes of oscillation is that it has been shown that higher modes can be more sensitive to material differences, particularly in the phase signal [132]. Garcia and co-workers have studied the theory of this type of imaging in several works [131, 133, 134] and explain that while the phase shift of the first fundamental frequency is sensitive to energy loss, the higher harmonics can be sensitive to tip–sample interactions that conserve energy as well, explaining the contrast improvement in higher harmonic phase imaging [134]. In recent years, more reports have emerged also giving further experimental evidence for the high material sensitivity of the phase shift at high harmonics [135–137]. This high sensitivity of the technique has been used to obtain high-resolution images in IC-AFM even of very soft samples [138, 139]. These materials require very low force imaging in IC-AFM mode to avoid damage, which reduced the contrast in the fundamental mode to the point where no sub-molecular details were visible, but increased details were available in the higher oscillation modes. In addition, it has been reported that using higher harmonics for feedback can improve imaging due to higher Q of the higher modes [135].

3.2   Non-topographic modes

Ever since the early papers on STM, scanning probe microscopes have been used to obtain more than just topographic information. In those early experiments, the first reports of a scanning-tunnelling spectroscopy (STS) experiments were made [140, 141], which consists of ramping the tunnelling voltage and monitoring the tunnelling current with the tip held fixed over a particular part of the sample surface. The use of the word 'spectroscopy' has continued into the field of AFM, where 'spectroscopic' techniques are different from 'microscopy' techniques in that they probe properties of the sample other than topography. The most well-known example is probably force spectroscopy.

3.2.1   Force spectroscopy

Force spectroscopy involves maintaining the x-y position of the AFM probe fixed, while ramping it in the z axis, to measure the deflection as the tip approaches and retracts from the sample surface. As such, force spectroscopy consists of simply measuring force–distance curves, as shown in Figure 3.15. The great utility of this technique is that the AFM directly measures the force between the contacting atoms or molecules on the end of the probe and sample surface, and as the cantilever may be highly flexible, and deflection sensitivity with optical lever-based instruments is very high, single-molecule interaction studies are possible. Often, an AFM tip will be modified with grafted molecules of interest [142–145], although such experiments have also been reported with bare AFM tips [146, 147], colloidal probes [148–150] (e.g. silica spheres, which may be themselves chemically modified), and even micro-organisms [151, 152]. The surfaces probed have been of even wider variety. Again, for molecule–molecule interactions studies, often a flat substrate will have the molecules of interest grafted on [153], but also cell membranes [154], micro-organisms [155, 156], whole living cells [157] and a wide variety of solid surfaces including polymers [158–160], metals [161], ceramics [162] and more have been probed.

There are a number of experimental issues which must be taken account of in order to perform force spectroscopy. These include:

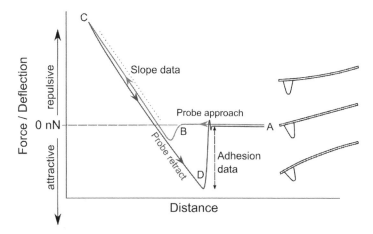

Fig. 3.15. A model force–distance curve. At point A, the probe is far from the surface, at B 'snap-in' occurs as attractive forces pull the probe onto the surface. The force becomes repulsive as the probe continues to be driven towards the sample. At some user-defined point C, the direction of travel reverses. At point D 'pull-off' occurs as the force applied to the cantilever overcomes tip–sample adhesion. Adhesion data is used for force spectroscopy while slope data is used for nanoindentation (Section 3.2.2).

(i) The number of interacting molecules. Depending on the tip radius, a large number of molecules are likely to be able to interact with the surface at one time.

(ii) Orientation and accessibility of interacting molecules. Typically, the investigator would like to make comparisons between the molecular interactions measured at the surface, and results from solution studies, but the grafting of molecules to the tip may affect the results.

(iii) The speed of approach and withdrawal of the tip for the surface will affect the results.

(iv) Experimental environment. One advantage of AFM is that it may be carried out in almost any environment. For most chemical and biological work it is useful to carry out the experiments in liquid. It is simple then to change the composition of the liquid to see how it affects the results. For example, to prove antibody/antigen interactions, commonly blocking antibodies are injected into solution, after which forces may disappear to zero [163].

(v) Statistical variation in results is typically very large. This means increased experimental time, which is not normally a problem, as each force curve typically takes less than 1 second to acquire, but in addition a very large dataset is typically generated, and a lot of data analysis is likely to be required.

In reality, the results from force spectroscopy between molecules rarely look much like the cartoon in Figure 3.15. Usually, specific forces between molecules lead to much more complicated results. An example is shown in Figure 3.16. In the blue (retract) curve, several typical features can be seen. One is the almost-flat region labelled a. In this region, polymer chains linking the molecules to the AFM tip were unfolding. During

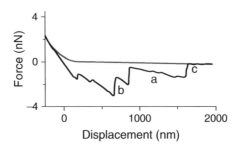

Fig. 3.16. An example of real force spectroscopy data: curves measured on *M. xanthus* cells. The red trace is the approach, and the blue is the retract curve. Reproduced with permission from [164]. Copyright 2005 National Academy of Sciences, USA.

unfolding, only very weak bonds are broken, so there are only small vertical deviations in the trace. At b, the probe applied sufficient force to break the bonds, as the molecule breaks away from the receptor. Note that at this point, a single vertical movement may be expected, but the step is staggered, indicating that multiple bonds are broken, and only at point c is the tip finally free of molecules linking it to the cell surface. In a case such as this, it is necessary to decide if the vertical distance (i.e. the force of adhesion), seen at point b, represents the adhesion of one molecule, that of two molecules, or of an unknown number. This is why it is difficult to automate data analysis in force spectroscopy, and this combined with the typical requirement to collect hundreds of data points, means data processing for such experiments can be very time-consuming. Some ways to improve the situation include reducing the chance of multiple interactions in the first place by for example spacing the grafted molecules out on the tip, or looking for multiples of single forces in the 'spectrum' of forces measured [144].

It can be useful to perform force spectroscopy in a grid-like pattern over the sample, leading to the possibility to locate specific chemical groups on a sample surface [146, 160, 165]. It is important, however, to remember that even highly specific measurements like adhesion–force interactions, may be affected by sample topography [159]. In this mode, force spectroscopy is sometimes termed chemical force microscopy [166]. A major application of force spectroscopy is protein unfolding, which uses the AFM force sensitivity to probe mechanical unfolding of large protein molecules, a biologically important process, which is covered in Section 7.3.5.1.

3.2.2 Nanoindentation

If instead of measuring the data as the AFM withdraws from the sample surface, we record the data measured as the tip contacts with and presses onto the sample surface, we are carrying out a different experiment, called nanoindentation. Another technique known as nanoindentation exists [167], which uses a dedicated machine to measure load–displacement curves as a hard indenter (for example diamond) presses into a sample. Typically, such instruments are designed to create a series of indents (holes) in a sample, and allow the measurement of the sizes of the indents (by, e.g. light microscopy), and are sensitive to forces in the micronewton range. By carrying out an analogous experiment using AFM we have some advantages and some disadvantages. These are summarized below.

Advantages of AFM-based nanoindentation

- High load sensitivity – load sensitivity may be as low as piconewton, although even for soft materials the required sensitivity is not likely to be greater than a nanonewton.
- Inbuilt ability to measure the indents created, at high resolution in x, y and z (see Figure 3.17).
- High positioning resolution – i.e. we can choose small regions of a sample, or perform the experiment on very small samples.

Disadvantages of AFM-based nanoindentation

- Non-perpendicular probe approach – quantitative nanoindentation requires the indenter to approach the sample perpendicularly, which is not the case normally for AFM. This problem can be overcome, with care.
- Non-linear z positioning. Unless the system is equipped with linearization in the z-axis this can cause some serious problems.
- The system must be calibrated to extract real forces.

For nanoindentation on hard materials it is necessary to use a very stiff cantilever and a hard probe. Typically, one might use a cantilever machined from steel, with a diamond tip glued to the end [168]. Such levers may be appropriate to perform nanoindentation and can be capable of imaging the sample, but typically give relatively low-resolution images; on the other hand, they are absolutely necessary to indent hard material such as metals. Many authors have also carried out nanoindentation with normal AFM probes [168–172], but it is necessary to characterize the tip radius and cantilever carefully for quantitative results. One advantage of such an approach is the ability to select from a wide range of spring constants; the highly stiff nanoindentation cantilevers previously referred to are inappropriate for soft samples. One common approach to simplify the problem of tip radius determination (see Chapter 2) for nanoindentation measurements is to use a colloidal probe, i.e. to use a normal AFM cantilever without a tip, but with a small spherical particle in its place [150, 173]. If nanoindentation experiments are carried out in a grid pattern over the sample surface, then it's possible to determine the spatial variation of hardness and softness [158, 174, 175]. Data analysis for nanoindentation is often made by modelling the indentation via the Hertz model, which requires knowledge of the shape of the tip, and assumes only elastic compressions of the sample take place [162, 176]. For more discussion of data treatment for nanoindentation see references [168, 176, 177].

Applicability

Despite the quantification issues associated with carrying out nanoindentation using AFM, it has been widely applied. It is particularly useful to look at relative hardness and softness. For example, it can give an idea about differences in hardness and softness in different parts of a sample With nanoindentation mapping, the measurements can be made quantitative, whereas for many other techniques such as phase imaging (see Section 3.2.3.2), it is hard to know if differences are due to mechanical or adhesive properties of the sample. Therefore nanoindentation has been commonly used to study heterogeneous materials such as polymer composites [158, 181, 182]. Furthermore, the high positioning accuracy means it's possible to look at small features not possible by traditional nanoindentation,

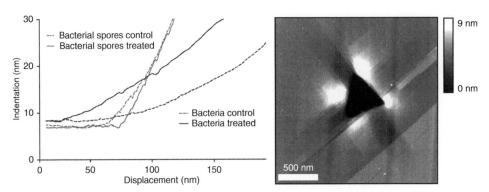

Fig. 3.17. Examples of nanoindentation measurements with the AFM. Left: force–distance curves measured with the AFM on individual bacteria. Black curves: typical data measured on untreated and treated *Bacillus* vegetative bacterial cells. Red curves: data measured on *Bacillus* spores. The data showed that the treatment made the cells softer, but the spores were much harder than the vegetative cells [178]. Right: AFM image of an indentation made by a dedicated nanoindenter. The indentation is in a magnesium oxide crystal, and the image shows the indentation (black triangle) pile-up – material pushed out of hole (white features at triangle corners), and also shows long-range dislocations in the crystal structure (diagonal discontinuities) [179]. Reproduced with permission from [180] and kind permission from Dr C. Tromas.

for example individual micro-organisms [169, 183] (see Figure 3.17), living cells [176, 184] or micro/nanoparticles [185–187]. Some more examples of applications of nanoindentation are given in Chapter 7.

3.2.3   Mechanical property imaging

Nanoindentation is a very useful technique for mechanical characterization because of the possibility to collect truly quantitative data on the mechanical resistance of samples. However it has several drawbacks, including the complicated data analysis, and its relatively slow data acquisition. The very low rate of data acquisition compared to normal imaging AFM modes is a major drawback. For an image with 512×512 data points, a full set of nanoindentation data would require many hours to collect, leading to problems with thermal drift of the sample. For this reason 'imaging' type studies with nanoindentation tend to be used only at very low resolutions (100×100 data points or less). One way to overcome this limitation is to measure the interaction of the probe with the sample surface while it acquires topographical data, and use this information to derive mechanical information about the sample surface. This has two advantages, firstly, data is acquired at a much faster rate, and secondly, the mechanical information collected may be correlated directly with the measure topography. There are a number of modes which acquire mechanical information about the sample surface in this way, and they are described in the following sections.

3.2.3.1   Lateral force microscopy

As described in Section 3.1.1, in contact mode, the vertical deflection of the cantilever, measured as the difference in signal between the top and the bottom of the split photodiode, is used as the feedback signal. However, if we compare the left- and right-hand sides of the split photodetector, we obtain the lateral deflection signal. When measuring this signal, the technique is sometimes called lateral force microscopy, or LFM. The reason why measuring this can be useful is that this signal contains information about the mechanical interaction of the probe tip with the sample surface. The lateral twisting of the cantilever is a measure of the friction encountered by the tip as it scans over the sample. Thus, this signal is sensitive to the nature (shape and frictional properties) of the surface. For this reason, LFM is sometimes also called friction force microscopy (FFM), and the lateral signal is sometimes referred to as the friction signal, although the signal obtained laterally contains more information than just the friction felt by the tip. It is important to bear in mind that the lateral bending is coupled with vertical bending of the tip, and contains information about the shape of the sample, as well as its material, because friction depends on the slope the tip is travelling along [77, 188]. However, using this technique it *is* possible to get quantitative information about variation in sample properties. Some examples of this are shown in Section 7.1.4. A discussion on calibration of lateral signals is included in Section 4.2.

As mentioned previously, it is not normally necessary to measure AFM height signals in more than one fast scanning direction. The situation in the case of the lateral deflection data is somewhat different. The lateral deflection signal will normally always be different in the two directions, as the cantilever will twist by a certain amount assuming there is some measurable lateral component to the tip–sample force (i.e. friction). Therefore, even on perfectly flat, homogeneous samples, the two images will be different from each other in the magnitude and possibly sign of the signal. In general, changes of slope will affect forwards and backwards scans oppositely, and changes in friction due to material contrast will give greater or smaller difference between the forward and reverse scans. This is shown schematically in Figure 3.18.

From Figure 3.18 it is possible to see that changes in slope and changes in material contrast have different effects upon the lateral deflection signal. If the user subtracts the left-to-right and right-to-left signals from each other, in the case of the slope change, the result will be a signal with almost no contrast. However, in the case of the material friction change, the resulting signal will be sensitive to the sample friction. Larger friction will give a greater difference between the forward and reverse scans, while lower friction will give a smaller difference. Thus, collecting both forward and reverse direction scans and subtracting them in LFM can give useful information [160, 189].

3.2.3.2 Phase imaging

'Phase imaging' in AFM refers to recording the phase shift signal in intermittent-contact AFM. In 1995 for the first time, the phase signal was described as being sensitive to variations in composition, adhesion, friction, viscoelasticity as well as other factors [190]. Then in 1996 Garcia and Tamayo suggested that the phase signal in soft materials is sensitive to viscoelastic properties and adhesion forces, with little participation by elastic properties [191]. It has been a common assumption ever since that phase contrast will show adhesion or viscoelastic properties [192, 193]. In fact, as shown in the examples of phase contrast in Figure 3.19, phase contrast from material properties is seen in a wide variety of samples, but also reflects topometric differences (differences in slope). This is because the phase is really a measure of the energy dissipation involved in the contact

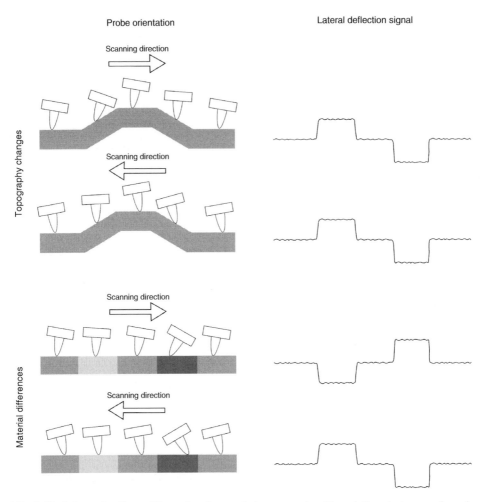

Fig. 3.18. Schematic of lateral force signals recorded on a sample with variations in topography only (top) and in material friction only (bottom). Darker colours represent material with higher friction. Note that in the case of topography changes (upper), the difference between the forward and back lateral deflection signals is constant; for material contrast (lower), the difference changes.

between the tip and the sample [194–196], which depends on a number of factors, including such features as viscoelasticity, adhesion and also contact area [197]. As contact area is dependent on the slope of the sample, the phase image also contains topographic contributions, so unambiguous interpretation of contrast in phase images is best left to flat samples. Even in such cases, understanding of the contribution of the individual factors to the phase shift is not trivial. For more details on this topic, the reader is recommended to read the excellent and comprehensive reviews by Garcia [108, 197]. Despite the complications involved in interpretation, phase contrast is one of the most commonly used techniques for 'mechanical' characterization of sample surfaces, probably due to the

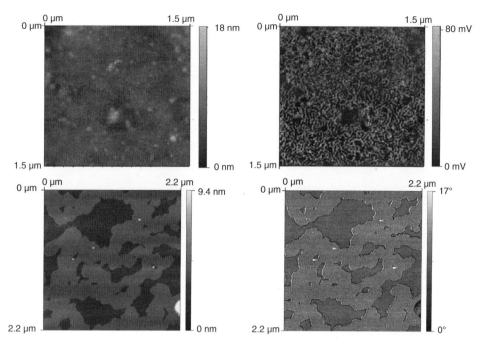

Fig. 3.19. Examples of phase contrast in IC-AFM on different samples. Top: a triblock copolymer topography (left) barely shows height differences for the different phases. The phase image (right) shows clear contrast. Bottom: Langmuir–Blodgett film on mica, the high topography region (the monolayer) has a higher phase contrast than the mica in the phase image. This image shows how the edges of these phases also show different contrast in the phase image, due to changes in tip–sample contact area.

popularity of IC-AFM, and the fact that obtaining the data is very simple and does not require post-processing of the data.

3.2.3.3 Other dynamic modes

A number of less commonly used oscillating modes have been reported [198, 199], these are typically variations on IC-AFM, designed to make simultaneous acquisition of sample properties and topography simpler or more quantitative. An example of this is jumping mode AFM [198, 200–204]. This is a variant of IC-AFM, the difference being that in jumping mode, the movement along the fast scan axis is discrete, rather than continuous, and the electronics are set up to record the cantilever deflection at specific points along the force–distance curve during each oscillation. The advantage of such a technique is that if, for instance, the points recorded are equivalent to points a and b in Figure 3.2, the tip–sample adhesion may be obtained, or slope data (see Figure 3.15) could be recorded to qualitative sample stiffness. The advantage of this particular mode is that the relatively high-speed scanning of IC-AFM can be combined with the acquisition of such data. This is also the aim of pulsed-force mode [199, 205–207], which operates in a very similar way to jumping mode, although fast scan axis movement is continuous, like normal IC-AFM. As

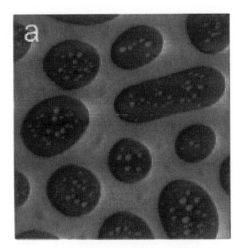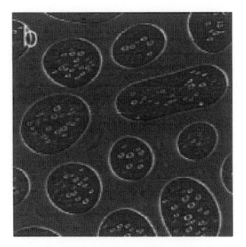

Fig. 3.20. Example of pulsed force mode. The sample is a polystyrene-polymethylmethacrylate blend. A: topography, B: adhesion, both measured simultaneously. Note the bright borders between the phases are due to increased tip–sample contact area, and the adhesion image is in agreement with that measured by force spectroscopy [159]. Reproduced from [206] with permission.

with jumping mode AFM, a major aim of pulsed-force AFM is to obtain adhesion data [208], but collection of other data points can again lead to sample stiffness data [199]. An example of the results from pulsed force mode is shown in Figure 3.20.

3.2.4   Magnetic force microscopy

The potential of using AFM to measure magnetic properties was realized quite early in the history of AFM [105, 209, 210]. Magnetic fields decay quickly with distance, so in order to measure local properties the probe must be very close to the surface, hence the applicability of AFM. The most typical experiment carried out is known as magnetic force microscopy (MFM) [211]. In this mode, the presence and distribution of magnetic fields is measured directly, by using a magnetic probe. Typically, these consist of standard silicon cantilevers with a thin magnetic coating. Typical materials used for the coating include cobalt, cobalt-nickel and cobalt-chromium [212]. The addition of such coatings can have two detrimental effects on the cantilever: firstly these materials are typically softer than the underlying silicon, and thus may increase wear rate, and secondly, any coating added to the end of the tip will increase the radius, and thus decrease the resolution of the experiment. Typically, magnetic forces are orders of magnitude lower than other tip–sample forces when in contact, and thus it is useful to measure them with the tip at a certain distance (of the order of 5–50 nm) from the surface, thus reducing the interference from short-range forces. This can be carried out in a number of ways [213], some of which are illustrated in Figure 3.21. These techniques all have some practical advantages and disadvantages, but are basically variations on a theme. In 'lifting'-type modes, the topography of the sample is measured first, followed by raising the probe, and scanning again to collect the magnetic data. One method is to collect a normal topography scan, and then change the z set-point to lift the probe from the surface and collect a 'magnetic image'

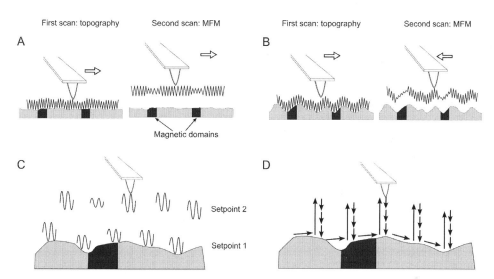

Fig. 3.21. Schematics of various implementations of MFM. A: lifting probe between topography and MFM images. B: Bard method of lifting lever between scan lines. C: z set-point oscillation. D: Hosaka method of moving probe close to surface, and recording MFM signal at various points for each height.

(Figure 3.21A). This works well for flat samples, but is prone to problems of features from the sample topography appearing in the MFM image, and also to problems from thermal drift. As described by Bard [214], an improved method is to record the sample topography first, then lift the probe, and measure the long-range forces while following the shape of the topography, but at a certain 'lift height'. This is applicable to STM, EFM (see the following section), or MFM. Typically, this is carried out in alternate scan lines, allowing the topography data to be included in the second, magnetic scan line, meaning the probe can stay approximately the same distance above the sample, even with changes in topography (Figure 3.21B) [215]. It's also possible to change the z set-point while scanning, meaning the probe will be constantly moving towards the sample to check the topography, and then away again to register magnetic field information (Figure 3.21C). Finally, in the method described by Hosaka [216], at each pixel the probe is lifted above the surface, and the field is measured at several points as the probe is lowered again (Figure 3.21D), to obtain a magnetic field gradient. The probe is then moved to the next lateral point, lifted again, and so on. This method is probably the least prone to thermal drift, but is rather slow to implement. Whichever method is used, lifting the tip from the surface reduces resolution, and resolution in MFM is typically no greater than 30 nm laterally [212].

For these lifting modes to work, it helps if there is little sample drift, or to have linearized scanners. Typically, MFM is carried out in one of the dynamic modes, and the magnetic effects on the cantilever are detected via phase shift, but they may also affect the oscillation amplitude. Unfortunately, even at lift heights of several tens of nanometres from the sample surface, short range forces other than magnetic interaction may affect the cantilever oscillation, giving a false indication of magnetic contrast [217], an effect which is sometimes overlooked. One way to overcome this problem is to carry out two scans with the cantilever magnetization orientation in opposite directions, and subtract them from each other. Non-magnetic forces should then cancel out, leaving typically a sigmoidally-shaped contrast in the lines scans where magnetic interaction took place [218]. An example image obtained in MFM via the Bard method is shown in Figure 3.22.

Although MFM is a relatively simple technique to obtain magnetic contrast at a high resolution, quantification of MFM signals is complicated, and when trying to measure the magnetic domains on a soft magnetic material, the domains on the probe can cause a change in the domain structure on the surface. Readers interested in more detail on the issues in quantification of MFM signals are directed to the work of Proksch *et al.* [218, 219]. It is worth pointing out here that there are a variety of other magnetic characterization techniques using the AFM, such as MRFM that involve considerably more equipment than a commercial AFM [220], so are outside of the scope of this book.

Applicability
The initial interest in the standard MFM technique grew largely because of the potential industrial applications. The data storage industry is largely based around creation of magnetic nanodomains of the size range of a few hundreds of nanometres, and there is no other technique to accurately measure such features. Therefore MFM has seen much use industrially, particularly in data storage applications [210, 213]. More recently, magnetic nanoparticles have become the focus of intense interest, and these are another field where MFM can be of great use [221]. The very small magnetic moment of the smallest particles can present a challenge, and much work has been carried out on particles of *ca.* 50–100 nm [221] but it should also be possible to examine the magnetic field from particles as small as 20 nm. Some more details of industrial applications of MFM are described in Chapter 7.

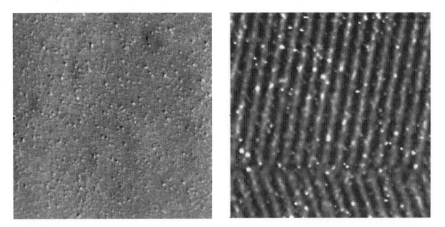

Fig. 3.22. Example MFM images. Left: topography of magnetic tape sample. Right: MFM image of the same region, showing magnetic fields above recorded data bits on the tape. Both are 10 μm × 10 μm images.

3.2.5 *Electric force microscopy and scanning Kelvin probe microscopy*

Electric force microscopy (EFM) refers to a technique analogous to MFM which enables the measurement of electrical fields with the AFM, rather than magnetic fields. Essentially, the technique can be applied by carrying out experiments in a lifting mode as described above, but without a magnetic coating on the cantilever. A standard silicon or silicon nitride cantilever may be used for simple EFM imaging, although conductive (metal-coated) tips are required for read/write applications, and more sophisticated electrical modes (see below). The equation for electrostatic forces between a probe and a surface having different potentials is given by:

$$F_{\text{electrostatic}} = -1/2(\text{V})^2 \frac{dC}{dz} \tag{3.3}$$

It can be seen that from Equation 3.3 and Equation 3.2 that the change in resonant frequency is proportional to the changes in capacitance as a function of the second derivative of z spacing. In other words, as long as there is a non-zero potential between the probe and surface, the frequency, and thus the amplitude and phase of oscillation will be sensitive to capacity of the surface.

EFM has been shown to detect trapped charge on surfaces [222], and in some cases gives clear contrast where none is visible in the topography signal. However, it has been reported that EFM is prone to topographic artefacts [223]. EFM, like MFM has the great advantage that it may be carried out with a standard AFM. A somewhat more sophisticated technique to measure tip–sample potential is scanning Kelvin probe microscopy (SKPM) [224, 225]. Figure 3.23 illustrates the portion of the SKPM instrument used for equilibrating the probe surface potential. The electronics used for mechanically vibrating the cantilever are not shown.

The principle of operation of SKPM is simple, that is when two surfaces have the same potentials, there will be no forces between them, so in Equation 3.3, $\Delta V = 0$. To implement the technique, a DC potential bias (V_{DC}) is applied to a conductive probe, which is further modulated by an AC signal (V_{AC}), so that

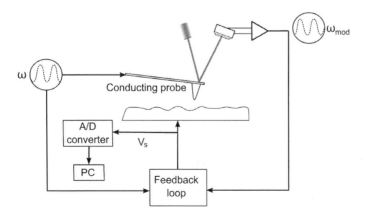

Fig. 3.23. Schematic illustration of instrumental set-up for scanning Kelvin probe microscopy.

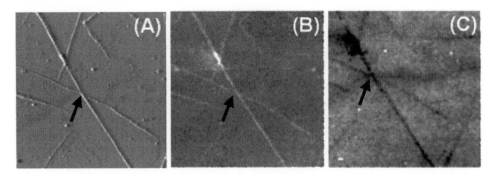

Fig. 3.24. Example of Kelvin probe and electric force microscopy. AFM height image (A, shaded image), Kelvin probe (B), and EFM (C) images of carbon nanotubes on a gold surface. The images are not all in exactly the same place; the red arrow highlights a connection between two nanotubes in each image. Reproduced from [227], with permission.

$$V_{\text{bias}} = V_{DC} + V_{AC} \sin \omega t \qquad (3.4)$$

In other words, the AC voltage is oscillating at the resonant frequency of the cantilever [224]. Thus, the probe's electric potential is varying at frequency ω. If the sample's potential is not the same, the difference in electrical potential will cause the cantilever to mechanically vibrate at the frequency ω, and which means that the electrical signal from the photodetector will be modulated at ω. A feedback circuit then compares ω with ω_{mod}, and outputs a DC voltage to the sample that minimizes the oscillation at ω_{mod}. This occurs when the applied potential V_{DC} is equivalent to the surface potential V_{s}. So the voltage V_{DC} that is require to minimize ω_{mod} is digitized with the A/D converter and displayed on the PC as the potential image [225, 226]. By SKPM, absolute values of the sample work function can be obtained if the tip is first calibrated against a reference sample of known work function.

3.2.6   Electrochemical AFM

Although not really a separate mode, it is worth mentioning that it is rather simple to study a surface as a function of applied potential using the AFM [228]. Changes in sample topography with applied potential are the results of electrochemical reactions, and so this technique is known as electrochemical force microscopy. *In situ* imaging of such processes is achieved with an electrochemical cell which is a modified liquid cell with the addition of electrodes to bias the sample and a potentiostat. By ramping the applied potential to the oxidation or reduction potential of the surface during scanning, or between scans, it is possible to directly observe oxidation or reduction processes on the sample surface. Such processes tend to give rise to small (or slow) changes in sample topography, hence the usefulness of electrochemical AFM. Furthermore, it is possible, using more modifications of the instrument, to combine imaging with electrochemical measurements at the nanoscale, a technique referred to as scanning electrochemical AFM [229]. An example image showing results from electrochemical AFM is shown in Figure 3.25.

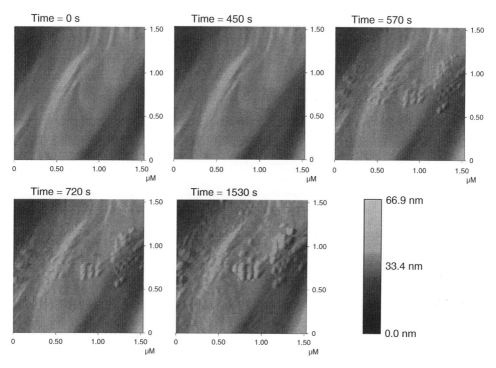

Fig. 3.25. Electrochemical AFM example. Images showing the morphology of a CdTe film during electrochemical deposition of Au, at various times (as shown in figure) at a potential of −0.35 V. Reproduced with permission from [230].

3.2.7 Thermal modes

It is possible to use derivatives of AFM to measure thermal properties of the sample [231]. Typically, this is done by using a resistive probe, which can locally heat the sample or measure the temperature locally, i.e. act as a thermometer. The first such probes were the so-called Wollaston wire probes, which consist of a very fine platinum wire bent into a v-shape. The apex of the v formed the tip of the probe. Later, micro-machined probes, developed from silicon nitride cantilevers, with a palladium layer which thins greatly at the tip apex, to act as the resistor, were developed. One common experiment involves applying a potential to the probe, which heats the resistance. As the sample is scanned (in contact mode), heat from the probe will flow into the sample, the amount depending on the thermal properties of the sample, and a feedback circuit adjusts the current flowing through the resistor, to keep the resistance, and thus the temperature, constant. Plotting the current applied to the probe gives the thermal image, and a topographical image is collected simultaneously. An example of the sort of data that may be collected with this technique is shown in Figure 3.26. This method is commonly termed scanning thermal microscopy (SThM). The thermal image in SThM is therefore a map of thermal conductivity, although it might be necessary to deconvolve topographic contributions [231]. By using temperature modulation (i.e. by supplying an AC current to the resistor rather than a DC current), the depth sensitivity may be changed, allowing for

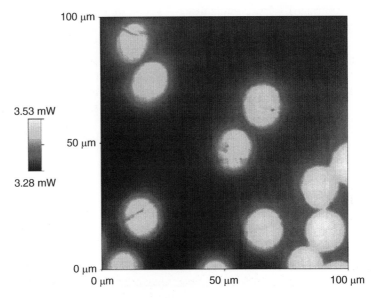

Fig. 3.26. Example of scanning thermal microscopy. Thermal conductivity image of a section from a glass filament/cyanate resin composite. The glass fibres clearly show greater thermal conductivity than the polymer matrix. Reproduced from [238] with permission.

discrimination of buried features [232]. This mode also allows for the imaging of heat capacity [231]. In addition to the imaging-type experiments, it is possible to perform many typical thermal analysis experiments using a similar set-up such as localized calorimetry or thermo-mechanical analysis [233–236]. The aim of all these techniques is to characterize materials thermally on the nanoscale. As such most of these experiments could be performed macroscopically on whole samples much more easily, so the main application is in heterogeneous materials. As well as specialized probes, SThM requires some simple external circuitry, and so its adoption as a standard AFM technique has not been widespread. However, such probes are commercially available, and the technique gives information not available by other means, so a large number of studies have been applied to polymer composites [237–239]; in addition, micro-organisms [231], pharmaceuticals [232, 236, 240], automotive coatings [241], metal alloys [242] and electronic devices [243] have been studied with SThM. The interested reader is directed to an excellent review for more information on this technique [231].

3.3 Surface modification

As well as measuring sample surfaces, an AFM may be used to manipulate or to modify the surfaces. The fine control of the probe motion over the surface makes even a standard AFM a versatile tool for manipulation surfaces at the nanoscale. There are a range of techniques that have been used to modify surfaces, notably including local oxidation [244], scratching [245] and dip-pen nanolithography [246].

Uncontrolled surface modification is usually an undesired feature of AFM, but it was realized early in the history of SPM that with care this technique had the potential to fabricate nanoscale devices [247]. One of the earliest of the nanolithographic techniques to be demonstrated was local oxidation [248]. In this technique, a bias is applied to the tip to cause contact potential difference while scanning the surface, resulting typically in an oxidation of the material at the sample surface. These experiments are commonly carried out on silicon and result in features of silicon oxide at the surface [244], although other oxidation-initiated reactions are possible [249, 250]. As noted previously, when scanning in contact mode, a liquid meniscus will be present between the tip and sample surface. In nano-oxidation this meniscus is vital because it provides the electrolyte for oxidation. Because of the importance of the liquid bridge for the reaction, the process is very sensitive to humidity, and the size of the meniscus has been reported as the factor controlling the smallest feature that it's possible to manufacture [244]. Local oxidation has been performed in contact [251–253], intermittent-contact [253], and non-contact mode [254]. If the tip is in the non-contact regime when the bias is applied, a capillary layer can spontaneously form, and it has been suggested that the water bridge under these circumstances is smaller than in contact mode, leading to smaller written features [254].

This technique has also been shown to be applicable to parallel fabrication [255–257], which is of great importance, because the main drawback of AFM-based nanolithography for fabrication is its slow speed [252]. Still, while local oxidation has been used to create nanoscopic functioning electronic devices [258, 259], fabrication of industrially useful structures on a large scale by this technique has yet to be demonstrated, even using parallel writing techniques.

To carry out surface modification with scratching techniques is a very simple technique, and is often used as a proof of principle experiment for lithography applications, because it is simple to apply to a range of materials. Structures have been built in polymers, silicon, metals and more by scratching [245, 249]. In principle, all that is required is to apply a high normal force to the sample, and use the lithographic controls in the AFM control software to direct the tip in the desired pattern. In this way, highly intricate patterns can be formed with this technique. Unfortunately, unlike oxidation or DPN, it is rarely applied to build structures with chemically different features, so the number of useful applications is relatively low.

Dip-pen nanolithography was invented in 1999 by Mirkin and coworkers [260], and has been shown to be a highly versatile technique. The great advantage of this technique is that almost any material that can be deposited on a surface can be used and formed into nanometre-scale patterns, although typically water-soluble molecules or very small particles are applied [246]. The idea is analogous to that of a macroscopic pen. The AFM tip is immersed, or dipped into a solution of the molecule to be grafted. With a hydrophilic tip, and aqueous solution, the AFM probe will become coated in a thin layer of the writing solution. Then, when the tip is in contact with the substrate, the grafting molecules are applied to the surface via the water capillary layer [260]. A schematic illustrating this is shown in Figure 3.27.

Like nano-oxidation, the size of the water bridge is a controlling factor in the dimension of the written features, as well as such factors as set-point, scanning speed, diffusion of the molecules on the surface, and tip radius [249, 261]. Examples of the sort of features that may be produced are shown in Figure 3.28. A great variety of 'inks' have been used, and

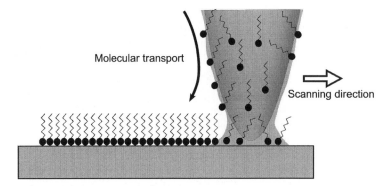

Fig. 3.27. Schematic of dip-pen nanolithography, showing how the water meniscus is used to transport molecules to the surface. Adapted from [260].

patterns have been created from organic molecules [260], proteins [262, 263], synthetic peptides [264], DNA [246], polymers [263], inorganic nanoparticles [265] and more [246, 249]. A major application of this sort of technology is in creation of arrays of receptors for parallel testing, e.g. proteomics, genomics, etc. For large scale parallel arrays of differing features, specialized DPN instruments, rather than commercial AFMS are typically used.

A number of other, less commonly used methods exist to modify surfaces with AFM [249, 266]. These include thermomechanical writing, which like SThM uses a resistance in the probe to control the temperature at the tip [267]. However, the temperature is used to modify the sample surface, rather than to probe it, and the high temperature is typically used to make holes in polymer surfaces without risk of damaging the tip. This has been investigated as a high-density data storage technique, and via the use of parallel probes (the so-called 'millipede' device [268]), has been shown to be capable of extremely high storage density [269]. Several authors have reported the use of the AFM to directly manipulate individual particles [270], molecules [271] and even atoms [272, 273] on a surface by for example, pushing, lifting and dropping or cutting [249]. These procedures are interesting for fundamental studies but are too slow to be of value as manufacturing techniques. Some examples of assembly using AFM are shown in Section 7.2.3. Finally, a

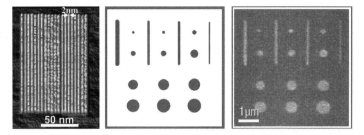

Fig. 3.28. Examples of AFM-based lithography. Left: polymeric patterns on silicon formed by anodic oxidation, showing line widths of approximately 2 nm. Reproduced with permission from [250]. Centre: a bit-map image used as the input for a dip-pen nanolithography (DPN) routine. Right: AFM (lateral force) image of the resulting surface patterns.

technique called nanografting is a variant of dip-pen nanolithography [274, 275]. It has the same advantage of flexibility – a wide variety of molecules may be applied to the surface [274, 276]. The difference is that it involves using the AFM tip to remove molecules from a previously modified surface, so that the molecules of interest, which are in solution, can form patches within the previous layer [277]. This has the advantage of leaving the molecules of interest surrounded with a potentially inert passivating layer covering the (typically) metallic substrate, making it useful for example fabrication of devices for binding studies [262].

Chapter 4

Measuring AFM images

Like all techniques, AFM requires some skill and practice to operate well, but learning to measure an AFM image is quite easy, and usually just takes a few hours of instruction and practice. Preparing the samples, setting up the instrument and scanning two to three images can take only half an hour. However, if it is an unknown sample that was never scanned by AFM before, it can take substantially more time to acquire useful images. In this chapter we discuss the procedures that can make measuring AFM images easier. This section does not replace the AFM manufacturer's user manual. Details specific to each instrument can be found in those documents. Instead, here we show the overall steps required for scanning a range of common samples, under typical conditions, and how to optimize conditions to get the best images. This chapter covers the most common imaging procedures; it focuses on contact mode and intermittent contact-mode AFM (IC-AFM). Non-contact-mode AFM is currently used much less widely than IC-AFM, so it is not explicitly covered here, but the imaging procedure is quite similar to that of IC-AFM. In addition to imaging procedures, some details on obtaining force–distance curves are included, as many users will also measure these. Figure 4.1 shows the major steps involved in measuring an image in an optical lever-based AFM.

4.1 Sample preparation for AFM

In general, sample preparation for AFM is very simple. For example, there is no need for the sample to be coated, electrically grounded, stained, or to be transparent, as required for some electron microscopic techniques. Some samples, such as thin films, can require no sample preparation at all. Other samples, such as human cells, or very small nanoparticles, may require considerable care in preparation for the best results. The 'rules' for preparation of samples for contact-mode AFM can be summarized as follows:

- The sample must be fixed to a surface. AFM is a surface technique, so all samples require some kind of substrate. Some common substrates for AFM are discussed below. If the sample consists of, or includes loose particles, these must be adhered to the surface before scanning. If some material on the surface is not well fixed down, it can lead to the AFM tip moving the material around on the sample surface. This can lead to a 'sweeping' of the surface, eventually clearing the substrate, with the particles being moved to the edge of the scan range. This sort of behaviour is particularly common in contact-mode AFM, as the tip never leaves the surface, and it can apply considerable lateral forces to the surface. Even if the sample is not 'swept' in this way, moving material on the surface will lead to inconsistent images, and 'streaking' as the tip encounters particles that are loose on the surface. It is also common for such particles to be transferred from the surface to the tip under these conditions. This will

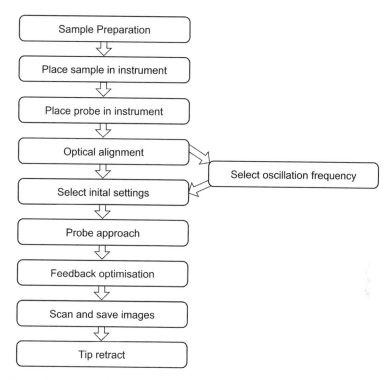

Fig. 4.1. The major steps involved in measuring AFM images. The oscillation frequency only needs to be selected for non-contact or intermittent-contact AFM.

lead to further inconsistency in the images, and it is also possible to permanently contaminate the tip, leading to strange artefacts in the images (see Section 6.1).

- The sample must be clean. Contamination in the form of particles or dried salts will make the underlying structure very hard to discern. Salt layers in particular are hard to discern optically, so that to the eyes the sample appears clean, but the salt layer will prevent imaging of the sample by AFM completely. Most samples imaged in air typically are coated with what in AFM is known as the 'contamination layer'. This liquid layer can be a mixture of water and hydrocarbons. Depending on the method used to image the sample, a light contamination layer (a few nanometres) may not prevent imaging of the underlying surface (see Sections 3.1 and 3.2). A thick (>50 nm) contamination layer can cause great difficult in imaging the underlying sample. Any particulate contamination will be imaged along with the sample, and complicate analysis. AFM tends to image *everything* on the sample, so it is important to remove as much contamination as possible.
- The features on the sample surface sample must be small enough to scan. AFM is a high-resolution technique, and most instruments are designed for small samples. The very largest scan ranges are on the order of 150 μm $\times$ 150 μm in x and y, and 28 μm in z, but a more typical configuration is a maximum range of 100 μm by 100 μm or less in x and y, and z limited to less than 10 μm. This is the size of the largest scan that

can be made, but most AFM instruments also limit the size of the sample that can fit into the sample (sample-scanning instruments are particularly limited). Specific instruments which allow scanning of very large samples do exist, however, they will typically include automated sample/head movement to allow for scanning of various areas across a large sample. Such instruments are typically aimed at industrial applications, e.g. scanning of whole semiconductor wafers.

- The sample has to be rigidly mounted into the AFM sample stage. A sample that is not well fixed down will tend to move while scanning, leading to distortion in the image. Vibration of the sample can also add noise to the image. The most common sample mounting for AFM is using a mounting disk made of magnetic stainless steel. This has the sample glued to it, sometimes using epoxy adhesive, which is highly rigid once cured, although double-sided adhesive tabs are also popular for less demanding applications. The magnetic disk is placed in the sample holder, which has a magnet in the centre. This arrangement keeps the sample very stable, and greatly reduces sample movement and vibration. Alternative arrangements where it is undesirable to use a magnet under the sample (e.g. for magnetic modes, or for optical access to the sample from below), usually involve some sort of sprung clips to securely hold down the sample.

Specific sample preparation techniques

The number of different types of samples that can be scanned by AFM precludes describing each one here, but it is possible to give some tips on preparing some of the most commonly examined samples here.

Particulate samples

Micro- and nanoparticles of all imaginable geometries and materials are very common samples for AFM, and imaging of a very wide range of different particles has been widely described [217, 278–284]. Often such samples come as an aqueous dispersion. The first step is to ensure the sample is as clean as possible, especially if the particles are very small (where the effect of contaminants is greater in relative terms). Where the dispersion is known to be very concentrated it should be then diluted. Often the ideal image will feature dispersed particles, so that the dimensions of the individual colloids can be measured. If the sample is to be imaged in air, then the sample is simply deposited by dropping a known volume onto a flat substrate and allowed to dry. Although AFM can operate either in air or liquid environments, imaging a sample that still retains significant amounts of water in air can be problematic, therefore improved imaging after drying samples thoroughly is common [285]. Often drying small (<100 nm) particles onto a flat surface is enough to 'fix' then adequately for AFM analysis, especially for examination by either IC-AFM or NC-AFM. For contact mode, especially for larger particles, such a procedure might not adhere the particles well enough to the surface. In this case, it might be necessary to have some chemical fixing to the surface, or use a special substrate (see below) [286].

Two factors in the sample preparation method that can have dramatic effects on the quality of results obtained are the solvent used to disperse the particles, and the substrate used. Water is generally the solvent of choice for such applications, as it is convenient, non-toxic and there are a wide range of methods available to produce highly pure water. It is always worth remembering that not all water is the same, however, so for samples with very small z-heights (e.g. nanoparticles <20 nm, proteins, nucleic acids), very pure water is required, in order that the contaminants do not mask the sample. Examples of the effects of differing water grades are shown in Figure 4.2. If very clean solution cannot be found, it may be advantageous to find a way to adhere the particles to the surface, which could be via silane or polycation modification [287–289], followed by washing, although this has the disadvantage of increasing substrate roughness [290]. Substrates for AFM are discussed in Section 4.1.1.

If it is desired to image a dry powder without dissolving in a liquid, a number of techniques have been described to immobilize particles. One technique is to immobilize large particles in a filter or similar porous substrate [291]. This can lead to the particles being sufficiently fixed to be able to scan them, and the top of the particle will be available to scan, but the full height of the particles will not be measurable. An alternative is to scatter the powder on an adhesive surface, such as a flat substrate with a thin layer of glue. Ideally the glue will be cross-linked/dried after the powder is applied, to avoid contamination of the AFM tip. Other systems that can work well with such samples include poly-l-lysine coated glass, and thin layers of wax, to which the sample is applied while the wax is soft (at elevated temperatures), and which solidifies on cooling [286]. For very small particles (<20 nm), many chemical modifications of the substrate surface produce a surface that is too rough for quantitative measurements. In such cases, deposition from ultra-pure water onto mica or HOPG is the best technique. Alternatively, some mica treatments have been described that increase the roughness only slightly [290].

Polymers

Polymer samples come in a wide variety of forms. Solid samples may require no preparation other than cutting to size and cleaning. Preparation of polymer films for AFM is also simple, and may be done by casting, spin coating, spreading, self-assembly, dip coating etc [146, 292, 293]. Typically such films are deposited on glass slides, as there is no requirement for very flat substrates.

Fig. 4.2. Example of the importance of clean solvent: images of a very flat substrate (mica) after deposition of drops of 'pure' water, followed by drying. Left: tap water. Middle: deionized, filtered water. Right: commercial ultra-pure water. All images are 2 μm × 2 μm, z-scale 4 nm.

Biomolecules – DNA and proteins

Oligonucleotides, especially DNA are very popular samples for biological AFM studies. DNA is usually deposited on mica [294], although HOPG has also been used [295, 296]. The negative charge of as-cleaved mica is a disadvantage for this application, as DNA Is also usually negatively charged. Typically this is overcome by treatment of the mica with a solution containing divalent cations, or deposition or imaging of the DNA in a solution containing such cations (typically $MgCl_2$ or $NiCl_2$ containing buffers) [297–299]. Alternatively, a procedure to bind oligonucleotides to mica with the aminosilane APTES has been thoroughly described [300]. The procedure should be followed carefully so that this does not increase the roughness of the surface. As with oligonucleotides, mica is the most commonly used substrate for protein absorption [294, 301, 302], but HOPG can also be used [303]. Again, divalent cations are commonly used to encourage protein binding to mica, if the proteins are negatively charged [301, 304]. The presence of monovalent cations in the buffer solution can compete with the divalent cations, and prevent the adhesion of a number of proteins to mica [305]. Other methods to fix proteins onto mica include covalent binding, although this may change the protein structure somewhat [306], and for membrane proteins, insertion into a lipid layer is a very suitable strategy [307, 308].

Cell cultures

Cultured cells are typically grown on some sort of support such as a Petri dish or glass microscope slide [309], to be directly mounted into the AFM. Where the instrument does not support such large substrates, microscope slides may be simply cut to size, or small cover slips used [310]. Cells may be fixed and dried before analysis or imaged *in situ* either in cell culture medium, or in a filtered buffer solution. In combination with temperature control, such a preparation can lead to the ability to image live cells [309–311].

Bacteria

Bacterial cells are common samples for AFM, see Section 7.3.2. Typically for imaging in air, bacteria are transferred to a clean buffer, dried onto a surface and extensively washed [169, 312]. For imaging in liquid, several procedures to adhere the cells to the substrate have been described [302, 313]. Without these treatments, the cells will normally be removed by the probe while scanning in liquid. Immobilization strategies include the use of gelatin coated mica to mechanically trap bacteria on the surface [6]. This has the advantage of not inducing chemical changes in the cells, as could be the case for binding the cells to the substrate with poly-l-lysine or other chemical treatments [184, 314, 315]. Another technique that might reduce the changes caused to the bacteria is allowing the formation of a biofilm on the substrate surface [184, 316]. For those bacteria which do form biofilms, this is the best way to adhere them to a surface for imaging in liquid. For spherical cells, physical immobilization in a solid substrate with holes (such as a membrane or lithographically patterned surface) has been reported to be very successful, although this is not appropriate for rod-shaped bacteria [317].

Nanotubes

Carbon nanotubes, nanowires and whiskers are a subset of nanoparticles. These particles are normally produced in large quantities as powders or are grown directly on a substrate.

Typically one of two methods is used for preparing nanotube samples for AFM imaging: catalyst growth or deposition. Catalyst growth is the best method for creating a clean sample for studying the unique properties of single-wall nanotubes. When preparing carbon nanotube samples for AFM imaging with deposition, it is important to use a dispersant. Very diluted dispersant suspensions of carbon nanotubes are spin coated on a silicon wafer or other flat substrate, rinsed thoroughly with water, and then dried in air. Any commercial spin-coater may be used.

Other solid samples

Metals or other solid samples may be imaged with little or no sample preparation. Cleaning (especially degreasing) can be required for some samples, and large samples may need to be cut to size. The lack of sample preparation for most solid samples is a great advantage of AFM, and means overall imaging speed with such samples can be higher than for electron microscopy. As with other microscopy techniques, polishing is required in order to observe metal grains [318].

4.1.1   Substrates for AFM

When preparing samples for AFM, especially particulate samples, a substrate must be chosen on which to mount the samples. In the case of very large samples, or very concentrated preparations, the nature of the substrate can be unimportant, but for many cases, it is crucial for correct sample preparation and good results. This is particularly important for high-resolution work, and looking at individual molecules in particular, for which an atomically flat substrate is usually required. For imaging of larger features, a substrate with a higher roughness can be adequate. As well as the roughness, the chemical nature of the substrate can be important. The intrinsic nature of the substrate is important in determining whether particular samples adhere well, and in addition, if substrate treatment is required some substrates are easier to modify than others. For example, highly oriented pyrolitic graphite (HOPG) is a commonly used substrate that is very simple to obtain in atomic flatness. This is because, along with mica, it is a layered material that is easily cleaved. Cleaving such materials exposes atomically flat faces, completely free of contamination. However, chemical modification of HOPG is not simple. On the other hand, gold is a highly stable material that is extremely simple to modify chemically, but while it is possible to produce extremely flat surfaces with it, it is considerably more difficult than for HOPG or mica. Table 4.1 summarizes some properties of commonly used substrates for AFM.

4.2   Measuring AFM images in contact mode

As shown in Figure 4.1, after sample preparation and placing the sample in the instrument, the next step is to insert a probe into the AFM. It is possible that a probe will already be inserted, but when beginning with a new set of experiments, a new probe is usually inserted. Great care must be taken when handling the cantilever chips as they are very small and very delicate. A clean pair of tweezers with flat tips helps. Usually the probe holder will have a slot for the chip and have some sort of small spring or clip to hold it in

Table 4.1. Properties of some commonly used substrates for AFM.

Material	Preparation	Roughness	Common samples	Notes[†]
Mica	cleaving	< Å (atomically flat)	All, especially single molecules [319, 320]	Cleaved material, so very stable in storage. Hydrophilic [294, 305]
HOPG	cleaving	< Å (atomically flat)	All, especially single molecules [295]	Cleaved material, so very stable in storage. Conductive. Hydrophobic [321]
Silicon	None[*] or only oxide removal	< Å to a few nm [322]	Lithography, electronic applications	Often the best choice for conducting applications [251, 256, 323]
Quartz or Glass slides	None[*]	1–10 nm	Larger samples or films, commonly used for cells [302]	Not especially flat but easy to work with and cheap [324]
Gold	Flame annealing or	< Å to a few nm	Chemically modified surfaces	Easy to chemically modify; large atomically flat terraces [325–327]
	template stripping	< Å to a few nm	Chemically modified surfaces	Easy to chemically modify; like cleaved materials highly stable in storage [328–330]

[*]Usually only cleaning is required.
[†]Use of each of these substrates has been described many times; representative references are given here.

place. To allow for manufacturing differences and the use of different length cantilevers, there is usually some room for manoeuvre in where you place the chip, usually just a few hundred micrometres. It is not possible to put the chip in the same place each time by hand, and even a few tens of micrometres will completely change the optical alignment required. It is sensible therefore to find a position that works and stick with it, in this way the realignment on changing the probe will be minimal. Some users find that placing the chip against the edges of the slot can give increased stability and more reproducibility of the chip position. Some alternative instrument designs use either pre-mounted cantilevers on larger 'cartridges', or alignment groves machined into the cantilever chips to help in placing the chip, but most rely on the manual insertion approach described above.

4.2.1 Optical alignment

After placing the probe in the instrument, the alignment of the optical lever is carried out. This is done in two stages. Firstly, the laser spot is adjusted onto the end of the cantilever. There will usually be two screws to adjust for this, one to move the laser parallel to the cantilever axis, and the other perpendicular. The exact procedure for the alignment can differ somewhat from instrument to instrument, depending on the view the user has of the cantilever and laser spot on the optical inspection scope. The user must take care not to look directly into the laser beam, as it can easily damage the eyes. Visualization of the laser spot can be done by placing a piece of white card or paper in the path of the laser. A general procedure for alignment of the laser beam is shown in Figure 4.3. Figure 4.3 shows the procedure for beam-shaped cantilever. For v-shaped cantilevers, the procedure is very similar but at steps 4–5 the laser spot is positioned between the cantilever legs.

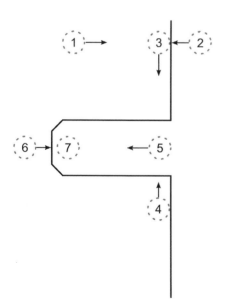

1 Laser spot starts in arbitrary place; move it towards chip.

2 Laser becomes occluded by cantilever chip. Move it back until it reappears.

3 Laser reappears; move it up or down towards lever, watch photodetector signal.

4 Laser passes lever; move back towards lever and optimise signal.

5 Laser is now in centre of lever close to chip; move it all the way to the left; watch signal.

6 Laser moves off end of lever; signal disappears. Move back onto lever, signal reappears.

7 Laser is now in the correct place; optimize photodetector signal once more.

Fig. 4.3. Laser alignment procedure.

Proper alignment of the laser is very important in order to obtain best results from the AFM. Poor alignment may reduce the sensitivity of the optical lever, could introduce imaging artefacts or prevent imaging altogether. For example, laser light spilling over the edge of the cantilever may reflect off the sample, and interfere with the laser light reflecting off the cantilever, see Section 6.6.2. The highest sensitivity is generally obtained with the laser spot centred over the position of the tip, that is, very close to the end of the cantilever, and in the centre (as shown in Figure 4.3 by point 7). One trick to check the laser is not on the edge of the cantilever is to observe the beam profile with white paper as described above; the edge of the cantilever will change the shape of the laser beam spot on the paper.

Having aligned the laser onto the cantilever correctly, it must then be correctly aligned with the photodetector. To do this the photodetector is translated until the laser spot is centrally located on the four segments, as shown in Figure 3.4. Sometimes there is a visual display of the photodetector in the AFM software, and sometimes just a numeric display of the signals from the photodetector segments. In both cases, the aim is the same, to get the laser spot to the centre of the detector, i.e. to equalize the signals from all four segments. This is a rather similar process to the laser alignment, and the only complication comes when the spot is completely off the detector, in which case the user might not know which way to turn the screws. If this is the case, the user simply turns the detector translation screw all the way in one direction, and then all the way in the other until the alignment is found, being careful not to apply too much pressure to the screws when the end of the movement is reached. In addition, often a third control is inserted in the optical path, which controls a mirror between the cantilever and the photodetector. This control directly rotates the mirror, and serves as a coarse adjustment for the photodetector alignment. This control is shown in Figure 4.4. In normal day-to-day operation of the AFM in air – for instance, when exchanging one probe with a similar one – the adjustment of this control is not required during the optical alignment procedure. There are two common reasons why the control might need to be adjusted. The most common reason for needing to adjust it is that when changing from air to liquid operation, the refraction of the laser at the liquid

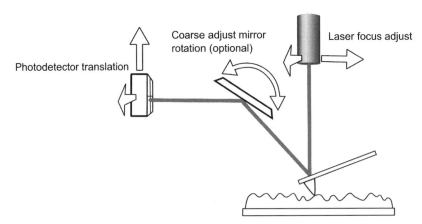

Fig. 4.4. Full optical system for optical lever AFMs, showing the various adjustments required for optical alignment (indicated by arrows).

cell's glass window/liquid interface considerably affects the alignment onto the photo-detector. Typically, the effect of this is that in one of the optical axes, the laser spot will move so far that the photodetector translation screws cannot move the detector far enough to account for this effect. In this case, a small adjustment of the mirror can correct for the refraction, and allow simple alignment. Because the misalignment by the refraction affects the optical alignment of the photodetector in only one axis, it is often useful to carry out the optical alignment in air first, and then add the liquid, followed by adjustment of the coarse control, particularly if viewing the cantilever when liquid fills the cell is more difficult. This makes adjustment of the coarse control far simpler, as a small change to this control changes the alignment drastically, and so it can be tricky to adjust without a prior alignment. The second situation in which the coarse adjust mirror might need to be adjusted is when a very large realignment of the photodetector is required because the laser spot is in a dramatically different position. This can be the case when changing from a short probe to a very long one or vice versa.

4.2.2   Select initial settings and probe approach

Once the probe and optical alignment are done, the next step is to initiate a probe approach. As described in Section 2.2.4 the woodpecker method is usually used for a safe approach to the sample, and to move into feedback. Depending on the instrument, the automated probe approach may be quite fast or quite slow. The relevance of this is that an instrument that approaches very quickly can be set to approach from a great distance, e.g. 1 millimetre, without taking too much time. Some instruments approach extremely slowly, and will take several minutes to approach a distance of only 100 micrometres. In this case, the probe must be moved close to the sample manually in order not to waste too much time waiting for the automatic approach. This must be done carefully in order to avoid uncontrolled tip–sample contact. The method to do this varies, but generally involves using the inspection microscope to alternately focus on the probe and sample surface in order to judge their distance from one another. More automated instruments can perform even this coarse approach procedure automatically. Before the automatic approach, the initial scanning parameters should be chosen, including scan size, scanning speed, gains, and set-point. For contact mode, the set-point is a measure of the deflection of the cantilever, and thus a measurement of the tip–sample force. However, the AFM instrument will typically show neither the true deflection (in nm) nor the force (nN), but the raw signal from the photodetector (in V or A). Thus, it can be somewhat difficult to know what initial set-point to use. It is best to use the smallest possible value as an initial step. However, during approach the actual deflection might vary somewhat due to thermal drift, long-range tip–sample interaction forces or other effects. If the set-point is too low, such variations will give rise to a 'false engage' where the instrument thinks the cantilever is on the surface, and the feedback is engaged, but the probe has not yet reached the surface. If suspected, false engage can be checked for by acquiring a force curve – if the curve is nothing like Figure 3.2, false engage is likely. Another way to check for a false engage is to watch the error signal (deflection signal) as the probe approaches the sample. A 'true' engage should show a 'jump' to the set-point the user chose. Slow, gradual movement towards the set-point is more likely to come from thermally induced bending of the cantilever. Thus, it's best to select a set-point somewhat greater than the cantilever deflection value, with some room for further deflection before the cantilever reaches the surface. The set-point may be further reduced if necessary once on the surface. Once initial parameters are chosen, and the probe is relatively close to the surface, an automated approach is carried out. Note that incorrect approach can easily damage a tip, an example of which is given in Figure 4.5. Some instruments allow adjustment of the automatic approach parameters, such as feedback values during approach, or approach speed. These should be changed only with caution, as the kind of damage shown in Figure 4.5 can easily result from using the wrong parameters.

4.2.3   Optimizing scan conditions

Optimizing the scanning parameters for the best possible image quality and most accurate images is probably the most important step in AFM data acquisition. Often 'standard' parameters are used initially for the approach, and such numbers might be provided by the instrument's manufacturer. However, these values will rarely, if ever, be suitable to obtained good images. The wide range of possible samples, scanning environments, and even probe manufacturing differences means that different parameters are used for nearly every scanning session. The method to optimize the parameters is an iterative one. The parameters are changed in steps, one at a time, until the tip is properly following the surface, and is giving a true image of the sample. Once the optimal parameters are determined, if the sample is homogeneous, and the instrument stable, the optimized parameters might be suitable for various images on the same sample. Changing to a similar sample with the same probe usually means small adjustments are necessary, again reached via an iterative procedure. Although it takes a while to fully master this procedure, following the method outlined in this chapter will allow optimization of scanning parameters in a few minutes.

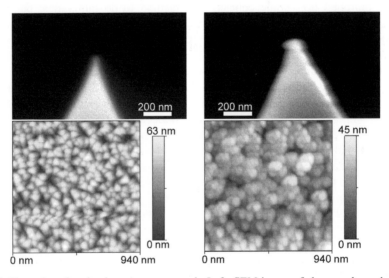

Fig. 4.5. Examples of probe damage on approach. Left: SEM image of sharp probe and an AFM image measured with the sharp probe. Right: SEM image of damaged probe and an AFM image measured with the damaged probe on the same sample.

When scanning the sample begins, it is useful to see a line-scan (a two-dimensional plot of the signal the instrument is recording). Often, the height data, as well as the z-error signal (in contact mode, the cantilever deflection) can be shown, and sometimes both forward and reverse signals are shown. The function of the AFM software that displays these signals is similar to an oscilloscope, so it is sometimes referred to as the oscilloscope window. This can be extremely useful for optimization of scanning. As forwards and backwards scanning lines measure (almost) the same parts of the sample, even when the slow scan axis is enabled, the two height traces should coincide. Large differences between forwards and backwards traces are an immediate indication that something is not right with the scanning. There are a number of possible reasons for forwards and backwards traces not matching but the most common reason is that imaging parameters (gains, set-point, and scanning speed) are not yet optimized. An example of the signals shown by the oscilloscope window is shown in Figure 4.6, illustrating the effect of different feedback settings on the results obtained on a simple sample. For clarity, only results from one direction are shown.

If the AFM probe is scanning over the sample in the normal raster motion (as shown in Figure 2.22), the features in the oscilloscope window will of course keep changing. It can be extremely helpful to have the probe scan in a line over the sample, without moving in the slow scan axis. Usually, the software will have an option to do this, and it is often the best way to adjust the scanning parameters as their effect on the scanning can be seen directly without interference from changes in sample topography. It is highly recommended that the line-scan option is used if difficulty arises in setting the gains, etc. After optimization, then the slow scan axis movement can be re-enabled and the image quality checked. When first learning to operate an AFM, it is helpful to scan a test sample and see the effect of the feedback parameters on the height (z voltage) and deflection signals. Such a sample has a very simple, reproducible topography (usually a series of square pits or posts), so it is easy to see when the scanning parameters are perfect. An image of such a sample, with the effect of varying the feedback parameters is shown in Figure 4.7. Some useful test structures are discussed in Appendix A.

The general procedure to use to adjust scanning parameters is as follows.

1. Increase feedback gains (PID values) step by step, observing for the start of feedback oscillation (the fine-structured noise seen towards the bottom of Figure 4.7). Typically, the integral gain is increased first, and then the proportional gain adjusted in approximately the same proportion.
2. When feedback oscillation occurs, reduce the gains again, until it disappears. The optimal value is the highest gain setting you can use without adding feedback noise to the image.
3. When gains are optimized, adjust the set-point. Ideally we would use the minimum value to keep the probe on the surface, in order to reduce probe wear. However, sometimes a greater force is required.
4. The gains may need to be optimized again to account for change in set-point.
5. Adjust scan speed if desired.
6. Gains and set-point may need adjusting once more to take account of change in scanning speed.

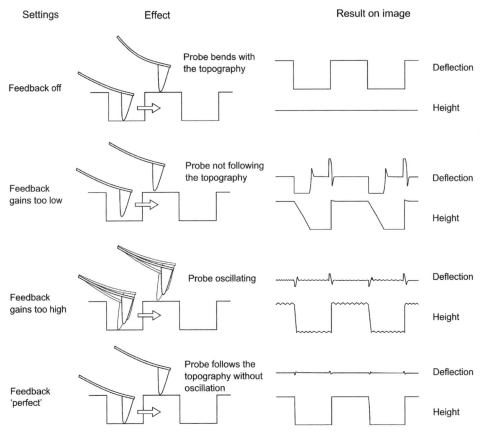

Fig 4.6. The effect of different feedback settings. The motion of the probe over a simple sample, resembling a calibration/test grid (left), and the oscilloscope window showing deflection (z-error) and height information (right). Note that feedback in AFM is never instantaneous, so the bottom example still shows some imperfections.

Note that gain, speed, and set-point are all related. At low scan speeds, low gains, and low set-point (low applied force) may be adequate, but faster speeds typically require higher gains, and might require higher set-point.

4.2.4 Choosing scan size and zooming

If the sample is heterogeneous, and certain features must be scanned, normally it helps to start with a large scan of the area, and then zoom to the feature of interest (see Figure 4.8). Zooming directly into features with AFM works well for instruments with scan linearization (see Chapter 2). With non-linearized scanners, it is best to zoom in 'gradually', by zooming to no than less than 50% of the current image size at a time. Thus, several zooms may be required to find the region of interest.

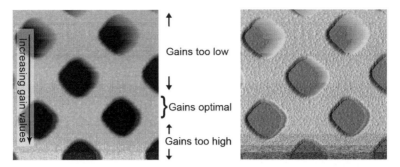

Fig. 4.7. Image of a test/calibration sample showing the effect of changing the gain settings during scanning. The height image is shown on the left, and the shaded height is on the right, which shows the fine details of the effects of the gain settings more clearly.

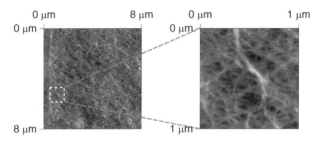

Fig. 4.8. Example of zooming to feature. Selecting a feature of interest in the left image gives the image at right.

4.2.5   *Other signals and measuring LFM images*

When scanning is optimized the user may choose which signal images to save. The user will always want to save a height image, whether from z scanner voltage or z sensor data, as it is the only signal with a fully calibrated z scale, from which the user can make height measurements. The error signal (cantilever vertical deflection signal) image can be useful to appreciate quickly the shape of the sample, as well as to spot areas where the height signal is invalid (areas of large or unchanging error signal). Some researchers publish the error signal; often it is a simple way to display features at different heights in the same image. The lateral deflection signal may or may not be useful. On many samples, the lateral deflection signal will show no more details than the vertical deflection signal. This is because lateral twisting of AFM cantilevers is much less sensitive than vertical bending [77]. However, if there is a requirement to record the lateral deflection (for example, if friction contrast is expected in the sample), it is simply a matter of selecting the signal to be saved. Unlike the other channels, for which forwards and backwards signals should be equivalent, it can be worthwhile recording both forward and backwards lateral deflection signals. This is because, as shown in Section 3.1.1, comparison of forward and backwards lateral force signals can help to distinguish topographical from frictional effects on the LFM signal. It is always worth remembering that tip–sample friction and thus lateral deflection will depend, among other factors, on the normal force applied by the tip to the sample, i.e. greater set-points will give greater contrast in the LFM. Some examples of LFM measurements are shown in Chapter 7 (Section 7.1.4), illustrating the difference between forward and backwards LFM signals. If the user wishes to obtain quantitative friction measurements from LFM, there are a number of calibration issues which must be addressed [331, 332]. While calibration of normal forces is an issue which potentially impacts on all AFM measurements, calibration of lateral forces is only important for quantitative LFM. Despite this, a large amount of work has been, and still is being done in order to understand how such a calibration can be made [331–336]. This is because the tip shape and radius, the cantilever twisting force constant, and the optical lever sensitivity must all be calibrated into order to fully understand the LFM signal. Also, unlike normal deflection there is no simple 'built-in' method to induce a lateral deflection of the cantilever in the instrument, making the optical lever calibration more complicated.

One of the first methods to be proposed for lateral force calibration, and probably the most widely used was described by Ogletree *et al.* in 1996 [77]. The Ogletree method (also known as the 'wedge' method) has a considerable advantage over some others in that it simultaneously calibrates the cantilever twisting constant and optical lever sensitivity. The method involves using calibration samples with known slopes to induce a fixed lateral force at the tip–sample interface. By comparing lateral force signals in different directions and at different normal forces (deflection set-points), a lateral calibration factor which enables measuring the tip–sample friction force in newtons per volt can be obtained. This method, along with improved versions using simpler materials has been widely discussed in the literature [331, 333, 337]. Other methods to calibrate the lateral friction constant include pushing the cantilever against a piezoelectric sensor [335], measuring static friction [336], quantitative comparison with a similar lever that's pressed against a side-wall while the bending measured [331], numerical methods [61] and others [338, 339].

4.3 Measuring AFM images in oscillating modes

Measuring images in oscillating modes is in general very similar to measuring images in contact mode, with just a few differences. Firstly a non-contact/intermittent-contact probe is used, usually with a much higher spring constant, and higher resonant frequency. One practical consideration here is that IC-AFM probes are even more fragile and easy to break than contact probes. A contact probe can sometimes survive a crash into the sample, as they are very flexible, but intermittent-contact probes nearly always break when this happens so even more care must be taken with them.

The optical alignment procedure is identical for the two techniques. However, once the intermittent-contact probe is loaded and aligned, the operating frequency must be selected. This is sometimes done via an automated routine, but often it is manual. Automated routines will usually require that the user enter an upper and lower boundary for the possible resonance frequency, and will then assume that there will be one peak within that frequency. The automated routines cannot cope under certain conditions, so it is important that the user knows how to manually select the frequency. This is done via a 'cantilever tuning' window in the AFM software. This program sweeps the oscillation frequency of the driving piezo up and down over a fixed frequency range and displays the amplitude of oscillation at each point. The user should have some idea of the natural frequency of the cantilever, so the start and end of the range to test in inputted to this window. This information is supplied by the cantilever manufacturer, and typically covers quite a broad range (e.g. 200–400 kHz). Within this range, the cantilever's oscillation should be visible as a single, strong peak. The presence of multiple or misshapen peaks in the frequency spectrum is an indication that something is wrong. The probe could be damaged or not fixed correctly in the probe holder. Once the peak is located, typically the user should zoom into the relevant part of the frequency spectrum to visualize the peak more clearly. An example of the view of a cantilever tuning window is shown in Figure 4.9.

It can be seen that the instrument often shows not only oscillation amplitude versus frequency, but also oscillation phase versus frequency. As shown here, the phase changes $180°$ – being $90°$ out of phase at the amplitude maximum, the greatest slope in the phase curve coinciding with the maximum in the amplitude curve. The meaning of these plots is also illustrated in Figure 4.9. The resonant frequency represents the point at which the amplitude is maximum, while the phase of the oscillation of the cantilever matches the applied phase ($\theta = 0°$). The actual operating frequency is at the maximum of the amplitude, but the user usually chooses a frequency a little way off the maximum (on the

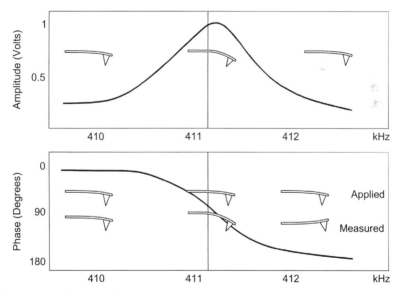

Fig. 4.9. Example of real amplitude and phase versus frequency plots used in cantilever tuning. The vertical lines represent the operating frequency, chosen by the user. Inset cartoons show the meaning of the amplitude and phase plots. Top: the cantilever's oscillation amplitude is maximized at the resonant frequency. Bottom: below resonant frequency, the measured cantilever oscillation follows the applied oscillation (phase, $\theta = 0°$), at resonance it lags the applied force ($\theta = 90°$), and above resonant frequency the applied oscillation lags measured oscillation further ($\theta = 180°$).

low-frequency size), to take account of the frequency shift as the tip approaches the sample. Selecting the wrong frequency (such as one at a higher frequency than the amplitude maximum) may allow imaging, but will usually give very poor images and possibly image artefacts. Having selected the operating frequency, the amplitude of the driving piezo oscillation is adjusted to give the desired oscillation amplitude of the cantilever. The oscillation amplitude, like the cantilever deflection, is normally shown only in terms of the photodetector output (e.g. rms amplitude in volts), so the desired signal varies from one instrument to another, but as discussed in Chapter 3, amplitudes in intermittent-contact AFM can vary from 1 to 100 nm [108]. In most AFM systems, an amplitude set-point is then chosen. For contact AFM, the set-point is a deflection value, which means that increasing the set-point leads to greater forces between the tip and the sample. However, for intermittent-contact mode, feedback is based on a *decrease* in amplitude, so a lower set-point means a greater tip–sample interaction force. For example, if the free oscillation amplitude, A_0 is 1.0 V, the user might choose 0.9 V as a conservative set-point, meaning the free amplitude will be allowed to decrease by 10% during approach at which point the system will go into feedback. Thus, a value of lower than 0.9 V would mean a greater force of interaction, and vice versa. Now, unlike contact-mode probes, IC-AFM probes are highly stiff, and so they are less prone to thermal noise and bending, and thus oscillation amplitude is highly stable. This should mean that false-engage is less of a problem. However, long-range forces between tip and sample do usually affect the oscillation slightly when operating in air. So, the user must once again be careful to avoid false-engage, so that it might be necessary to use a lower set-point than 90%. It is useful, again, to observe the error signal (the amplitude) as the tip approaches the sample, to help diagnose false engage. Once the oscillation frequency and amplitude set-point are chosen, an approach may be made. Due to the change in resonant frequency as the probe approaches the sample, it is sometimes helpful to withdraw the probe a little after a successful approach, and re-optimize the operating frequency.

Having approached successfully, scanning and optimization of gains are very similar to contact mode. So the procedure described in Section 4.2.3 can be used. The first-time user is reminded that the imaging mechanism for IC-AFM is completely different from that of contact-mode AFM (see Chapter 3). This means that optimal imaging parameters will usually be completely different for contact and IC-AFM, even on the same sample, and using similar probes. One aspect to be aware of is that the response of the probe to large topographic changes is rather slow in IC-AFM compared to contact mode, meaning scanning may need to be carried out more slowly. If the tip does not properly track the surface, either the speed may be decreased, the gains increased, or the amplitude set-point decreased. An example of the effect of scanning too quickly is shown in Chapter 6. Note that unlike in contact mode, it's not really possible to make force–distance curves in IC-AFM mode. One reason for this is that the cantilever is so stiff that trying to do this would apply a very large pressure to the tip of the probe, and damage it. However, often AFM systems do allow the user to obtain amplitude–distance curves. An example showing the utility of this is shown in Figure 4.10. Amplitude–distance curves have also found use in measuring long-distance forces on the tip, for example in MFM [340].

As described by Garcia *et al.* [341, 342], this sort of curve can serve a useful diagnostic purpose. As shown in Figure 4.10, it is possible to observe non-ideal curves, i.e. curves where there is more than one possible tip–sample distance at a particular amplitude

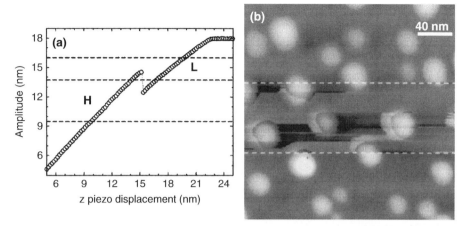

Fig. 4.10. Left: amplitude–distance curve showing jumping from a high amplitude state (H) to a low amplitude state. Right: effect of this jumping on an image. The dashed lines in the left-hand figure correspond to the set-points used in the three regions in the right images. With a set-point near the discontinuity in amplitude–distance, unstable imaging will occur. Reproduced with permission from[341]. Copyright 2000 by the American Physical Society.

set-point. This sort of situation will lead to instability in imaging. The origin of the instability in the image on the right would be unclear without the amplitude–distance curve. If such a feature is observed by the user, he should change the amplitude set-point to a value with a unique solution, shown by the upper and lower segments in the image in Figure 4.10. This example illustrates that sometimes scanning parameters can be adjusted in one of two directions in order to improve imaging. In the case shown above, the best solution might be to increase the set-point, rather than decrease it, as it will result in lower tip wear.

As described previously, in IC-AFM there are normally four types of signal that may be saved as images. There are signals of four types – the height (z piezo voltage signal, and z sensor, if available), amplitude (error signal), and phase signals. In addition, each channel may be obtained in one of two directions, or in both. It is up to the user which images to record and save. The height signals are the most important, as they are the only signals with a meaningful z scale, and the only signals from which we can make useful topographical measurements. It is not really necessary to collect signals in both directions, so only one height signal (typically the z sensor data, if it's available, otherwise the z piezo voltage) is normally collected. The amplitude signal can help in visualizing the shape of the sample, and in spotting features for later measurement in the height images. The phase image can serve a similar purpose, and in addition gives information about heterogeneity of the sample (see Section 3.2.3.2 for more discussion on this). Thus, the phase image can be highly useful on certain samples. It is rare that both forward and backward images are needed, so typically three images will be collected, height, amplitude and phase, either in forward or backward directions. If this is the case, it is important the user remembers to collect all images in the same direction, as forward and backward images may not be perfectly aligned with each other. If the phase image is of particular use in an application (e.g. for discrimination of phases in a polymer), it can be useful to optimize the phase signal. To do this, the amplitude set-point is usually varied, as a high set-point will generally give little contrast in the phase signal, while too low a set-point can damage or contaminate the tip, which will also negatively affect the phase signal.

4.3.1 Intermittent-contact mode in liquid

Imaging in IC-AFM in liquid is different from imaging in contact mode in liquid. Normal acoustic excitation of the cantilever in liquid leads to a number of peaks in the frequency spectrum, instead of the single sharp peak typically observed in air. The actual cantilever resonance is also shifted to lower frequency and is broadened (has reduced Q), compared to the response in air [128]. Finally, damping also reduces the amplitude of the oscillation, meaning that higher driving amplitudes will be required. The additional peaks arise from excitation of the liquid in the liquid cell, which further excite the cantilever [126, 127]. The shape of the cantilever's oscillation response in liquid will depend on the lever itself, the geometry of the fluid cell, and the distance of the lever from the sample [128, 343]. In consequence the user can be confused about which operating frequency to use, especially as the cantilever manufacturer will only specify the value of f_0 in air. However, many of these peaks, not necessarily near cantilever resonances, can be used to image the sample in IC-AFM, although some will work better than others [344]. Typically, best results will be obtained using the 'true' resonance, i.e. that obtained by direct excitation of the probe. Determining the frequency of this peak is sometimes a matter of trial and error. If the user does not know the typical frequency for a particular cantilever type, then it is best to choose a peak two to three times lower in frequency in the air peak which has a relatively sharp response. Try to image at the chosen frequency; if this does not work, try another, and so on. Once the direct excitation peak frequency is found, it is normal that a peak of similar frequency will exist for similar cantilevers. As the frequency of the cantilever resonances in liquid can be highly dependent on the distance between the lever and the sample, it is best to adjust the operating frequency when the probe is very close to the sample [345]. Commonly low-frequency (contact mode) cantilevers are used for IC-AFM in liquid, as samples are typically very soft when hydrated, and thus there's great potential for sample damage by IC-AFM in liquid [346]. There are many examples in the literature of imaging in liquid using IC-AFM as well as studies of oscillation of AFM levers in liquid, using a wide variety of probes, which can also help in determining the best oscillation frequency to use [344, 347–350]. It is worth pointing out that, as mentioned in previous chapters, alternative drive mechanisms exist which do not acoustically excite the cantilever, e.g. magnetic driving of the lever [124, 218]. Such direct-drive arrangements avoid the difficulties in choosing a peak to use – only the 'true' oscillation frequency will resonate. However, these arrangements make little, if any difference to image quality [125].

4.4 High-resolution imaging

Obtaining AFM images at relatively low resolution (scan sizes $>1\mu$m, resolution of >50 nm) is quite easy, but to obtain very high-resolution images (resolution on the order of 5 nm or less) can be considerably more demanding. To obtain very high resolution a large number of factors must be optimized.

The probe tip must be clean and particularly sharp. Even amongst probes rated as extra sharp, a large variation in actual tip radius is likely to be found as discussed in Chapter 2. For demanding applications, several tips could be tried, or a tip-check sample can be used. When all else fails, attempting to scan a well-known sample (especially one of the probe sharpness characterization samples) can often help to diagnose problems. Typically, if great results on such a tip-checker sample cannot be obtained, they won't be found from the sample of interest either. A list of samples suitable to characterize AFM probes is included in Appendix A.

The sample must be well fixed to the substrate, which should not be moving. The instrument must be at thermal equilibrium, and without drift. Sample drift is fairly easy to spot, and an illustrative example is given in Section 6.6.4. Sometimes the method used to fix the sample its substrate can be at fault, and a more rigid mounting (such as gluing with an epoxy adhesive) can help. Thermal drift characteristics can sometimes be helped by removing sources of illumination, which can heat the sample environment. Often, thermal drift is reduced with time, so leaving the instrument set up, with the laser aligned, the tip close to the sample, or in feedback with it, and the oscillation (if used) at the correct frequency, for 30 minutes to an hour, can reduce drift considerably.

Sources of external noise and the vibration isolation must be optimized. When scanning very flat samples at very high resolution, noise in the image that was previously invisible can often be seen in the image. In this case, the user must simply remove all possible sources of noise, such as lights or electronic equipment that are not required, and ensure the vibration isolation is fully functional, and uncompromised (e.g. by a mechanical connection from the stage to an un-isolated surface).

Finally, scanning parameters must be optimized. For very small scans it is possible to scan very quickly, as usually the feedback system does not have great changes in z-height to cope with. In addition, it is usually necessary to scan very quickly to overcome even small amounts of sample drift when imaging very small areas. For example, to obtain 'atomic lattice' resolution, with a scan size of ca. 10 nm, it is common to scan at about 60 lines per second. For high-resolution images, the perfect feedback is often found by making many tiny changes to the gains to reach the ideal imaging conditions.

4.5   Force curves

Force–distance curves are measured by monitoring the deflection of the cantilever as it approaches, touches, and withdraws from the sample. By default, therefore, they are measured in contact mode. Parameters to be selected by the user will include the x and y positions at which the curve is to be recorded, data density, movement speed (acquisition rate), maximum allowed deflection (force), length of the curve and more. If the area of interest is located in a particular region, it can be useful to image the sample before measuring curves. However, under some circumstances, such as when the cantilever is very stiff, or has been modified with a layer of molecules, it is not convenient to do imaging and force spectroscopy at the same time, especially imaging in contact mode which can damage a sensitive layer on the tip. So AFM software often has a separate mode for measuring force curves, which may also allow imaging in contact mode. Some instruments even allow a kind of hybrid IC-AFM/force spectroscopy, where imaging can be performed in IC-AFM, and when the area of interest is located, the instrument locates the surface using amplitude modulation, and only stops the tip oscillation during acquisition of a curve. This can reduce tip damage before tip acquisition. Attempting to measure force curves in selected regions of a sample with nanometre resolution can be challenging, partly due to sample drift, but also due to positioning difficulties and linearized scanners can help greatly. An alternative to carrying out force spectroscopy in one location is to perform the experiment in a grid pattern over the sample surface, thus enabling a grid of force curves which can be processed into a map of adhesion forces or sample stiffness. It's worth noting that at 1 Hz per force curve acquisition of 1 Hz, a 256×256 pixel map would take many hours to acquire, so such maps are usually obtained at low resolutions.

Regardless of the manner in which such a curve is recorded, the result is a plot of deflection versus distance, which the user usually wants to convert to force versus distance. The first step is to convert the deflection (V) into the actual distance the tip moved (m), then using the spring constant (N/m), and this can be converted to force (N). The normal deflection sensitivity is easily obtained by measuring the slope of a deflection signal versus vertical piezo displacement plot on a stiff, hard surface [142]. The surface chosen is often an extremely stiff one such as sapphire or stainless steel, but it is only important that it is considerably stiffer than the tip; measurements with flexible cantilevers could use any reasonably stiff surface for this. The user must obtain such a deflection-calibration curve to accompany each set of data without realignment of the laser on the cantilever; the exact alignment of the optical system directly affects this calibration [351]. Once this is obtained, the curve may be converted to force–distance with the normal spring constant. See Section 2.5 for procedures for calibration of normal spring constants.

Chapter 5

AFM image processing and analysis

AFM data needs to be processed, displayed and analysed to get the most out of it. AFM data has several advantages over data from other microscopes in terms of data analysis. Firstly, the data is in digital form, therefore there is no digitization step required. As AFM data is already calibrated, there is no need for a scale bar overlaid over the images because the scale is built into the data file. AFM data is by its nature three dimensional, so there are no calculations or guesswork involved in getting three-dimensional data. The height data is in the files and is easy to measure. Finally, AFM data often has different channels of information that are all recorded at exactly the same time, and in exactly the same position, removing the need for any 'alignment' steps. The data is saved by the AFM software as a file containing all the height and other data recorded by the instrument. Typically, the AFM software provided by the instrument's manufacturer has two main functions, image acquisition, and image processing. These could be separate programs, or may operate together. The latter part allows the user to open the file, and carry out the operations that will be described in this chapter. AFM data files are always proprietary formats; the data is not stored as simple image formats such as .tiff, .jpeg or .bmp which can be used by other microscope types. Such formats are not suitable for AFM images, because they cannot record the three-dimensional height data of AFM images, nor the additional data that can be obtained along with the height data (phase shift, cantilever deflection, friction, etc.). In addition to the actual data points, AFM files usually contain a lot of other information, such as scanning parameters, etc. The ADCs in AFMs usually record the data as 16 bit numbers (see Chapter 2), so the data must be stored also as 16 bit numbers to maintain the best resolution in the data. Optical microscope images and other photographs are usually stored as 8 bit numbers (256 shades per colour channel), because this is as many shades as the human eye can discriminate. Thus, these image formats are not adequate for AFM images, which usually contain far more information and resolution than can be simply visualized by eye, at least along the z (height) axis.

AFM file formats
AFM files vary widely in format from one instrument to another, but most formats are along the following lines. The start of the file will contain a large 'header' chunk, which may or may not be human-readable, but is easily read by AFM software. This is followed by a data section, containing the actual images. The header will normally contain some or all of the scanning parameters used to obtain the image (scan speed, gain settings, modes used, etc.). The most important function of the header is that it contains information allowing the second part of the file – the actual data – to be decoded. That is, it declares the scan size, and the 'conversion factor' which allows decoding from the 16 bit numbers in the data section into real data (distances, phase shift values, deflection values, etc.). It does not make sense to record all the data as real numbers, as this would reduce the precision available in the data. However, this storage method, combined with the large number of

possible formats and data types, means that manually reading data from AFM files is rather complicated. Additional information that might be included in the header include information about the instrument and software used for acquisition and descriptions of the sample, and even the probe used (these would be entered by the user during acquisition). There can be hundreds of lines of text stored in the header block of an AFM file. Following the header, the data section stores the individual data points as a simple block of consecutive 16 bit numbers (double bytes). The scan size and distances between these points are obtained from the information stored in the header.

AFM analysis software

As may be surmised from the brief description of AFM file formats above, writing software to open AFM files correctly is not a simple task. The best way for the user to analyse their data is always by opening the file in the analysis software provided by the manufacturer of the instrument on which the file was acquired. This will be the only analysis package that is sure to open and read the data correctly. There are many (more than 50) different AFM file formats, which are all mutually incompatible. Furthermore, many formats change over time, as the capabilities of the instruments are improved, so a program that can open old files from one format many not be compatible with the newer files. All this means that creating tools to read, manipulate and write these formats is far from trivial. However, some third party software does exist, and a list of third party software capable of opening and manipulating AFM data is included in Appendix C.

In this chapter, the operations carried out on AFM data are separated into processing, display, and analysis steps. The most important thing to remember when manipulating AFM data is to maintain the integrity of the original file as saved by the data acquisition software. The user should never save the AFM file over the original after changing the data in any way. The results of any operations which need to be saved or exported should be saved as new files–either new AFM data files, or as simple bit-map image files (.bmp, .jpeg, etc). All AFM analysis software has facilities to export to such bit-map files. One reason to avoid ever altering the original file is that AFM data files always contain more information than the user can see at one time. This can be in the form of alternate channels, more data resolution than is visible (there are more than 65,000 possible levels in the z-scale, whereas fewer than a hundred can be distinguished by humans), or in the form of the 'metadata' stored in the header block of the file. Furthermore, many of the processing steps that are referred to below cannot be undone. Thus, by processing, we can lose the ability to see the original data. Therefore, the user should always keep a backup copy of their original AFM data, and any processed files should be saved with a new name.

5.1 Processing AFM images

Processing steps change the AFM data; they include functions like filtering and background subtraction. Occasionally an image may not need any processing at all, but this is not common with most samples. All processing is done with the aim of clarifying the data already within the file. In other words, the purpose is to make it easier to measure and observe the features that have been measured.

5.1.1 Levelling

Levelling is described first here both because it is usually the first processing operation carried out on the data, and also because it is the most important, since it is the most widely used processing step applied to AFM data. In many cases, levelling is the *only* processing step carried out on the data, and it is required in nearly all cases. The reason levelling is required is that AFM images usually measure sample height. If the background in the image (such as the substrate on which the sample was deposited) has considerable tilt in it, the change in height of the background will mask the changes in height associated with the sample. AFM is often used to measure samples with very small heights, so even a small tilt in the sample background can have serious effects. Imagine a 20 μm × 20 μm image which contains some 50 nanometre nanoparticles, which are the features the user wishes to examine. If the substrate is tilted by only 1°, the height change from one side to the other of the substrate will be about 350 nm. This is more than enough to mask nearly all the nanoparticles. This concept, along with some examples of flattening operations, is shown in Figure 5.1.

In addition to tilting in the image caused by the AFM and sample not being perfectly orthogonal, a common problem in AFM images is scanner bow. An example of this artefact is also shown in Section 6.2.3. It occurs mainly in instruments that use tube scanners, and is caused by a swinging motion of the free end of the scanner. This leads to a curve in the image plane as shown in Figure 5.1. There are a number of different methods that can be used for image levelling, and these are discussed below.

Polynomial fitting

A very common method for levelling AFM images is by polynomial fitting, or 'line-by-line' levelling. In this routine, each line in the image is fit to a polynomial equation. Then, the polynomial shape is subtracted from the scan line, which leads to the line being not only flattened, but also shifted such that it centres on zero height. Typically each horizontal line of the image is processed in this way, although the process can also be

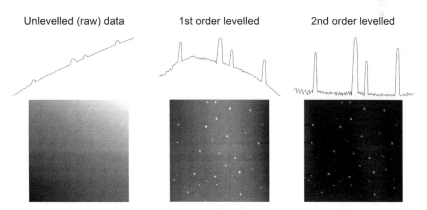

| Unlevelled (raw) data | 1st order levelled | 2nd order levelled |

Fig. 5.1. Illustration (top: line profiles; below: images) of the effect of first and second order polynomial line levelling. Left: unlevelled (raw) AFM image of nanoparticles showing tilt and scanner bow. Middle: effect of first order horizontal levelling – the nanoparticles are much clearer, but the curvature of the background due to scanner bow is still evident. Right: the effect of second order levelling with exclusions, the background is now flat.

Table 5.1. Effects of differing polynomial levelling orders.

Order	Effect
0	Only sets the height offset of each scan-line to the same value.
1	Fits a straight line equation to each scan-line, and does an offset.
2	Fits a quadratic equation to each scan-line, and does an offset.
3	Fits third order polynomial to each scan-line, and does an offset.

carried out on vertical lines. It is usually better to do this horizontally, because the horizontal axis is usually the fast scan axis. Therefore, any change in imaging conditions over the time of the scan will lead to horizontal discontinuities in the images, and will thus be well accounted for by a horizontal line-by-line levelling. The lines might then be moved up by a fixed amount such that the minimum of the image is at zero height, as it is a common convention not to use negative heights in AFM. The order of the polynomial equation can vary from 0 to 3 or more, as shown in Table 5.1.

For many images, a first order fit will suffice. If scanner bow is present in the image (often the case for larger scans), a second order fit will usually be appropriate to remove the artefact. Orders higher than third level are possible, but under these circumstances, the functions are likely to be fitting to real features of the sample, rather than just to the background. It's worth noting that line-by-line fitting procedures are particularly prone to causing levelling artefacts (see Section 6.3.1), and so feature exclusion (see below) is often required. Despite this, line-by-line, or polynomial fitting is often the most useful levelling procedure.

Two-dimensional plane fitting
This procedure tries to automatically fit a flat plane to the image, and subtracts the best-fit plane; it works well where the background is really flat, and does not include any curvature. The automatic routine will usually assume the entre image is to be fitted; this means that it can be subject to errors as large height features will reduce the accuracy of the fit to the background. Plane fitting does not introduce the kind of errors mentioned above that can plague line-by-line fitting. It is therefore a rather conservative levelling procedure, as although it's not very efficient with many images, it does not introduce any errors. For this reason, some acquisition software saves files with a plane fit by default, and some analysis software applies a plane fit whenever an AFM file is loaded. Usually, the user can turn off these 'autofitting' functions if necessary.

Three-point fitting
This procedure is similar to two-dimensional plane fitting, but is a rather more 'manual' approach. In this method, the AFM user identifies three points on the image. These points define a plane which is then subtracted from the image. The advantage of this method over the automatic plane removal routine is that if the user can distinguish the substrate from the sample features, he can ensure the three points used are on the substrate only, and this often leads to a better fit. However, because a flat plane is fitted, it is still only suitable for images with no curvature or scanner bow. It is particularly appropriate for samples with terraces. An example, showing the effect of different levelling algorithms on an image with a large terrace, is shown in Figure 5.2.

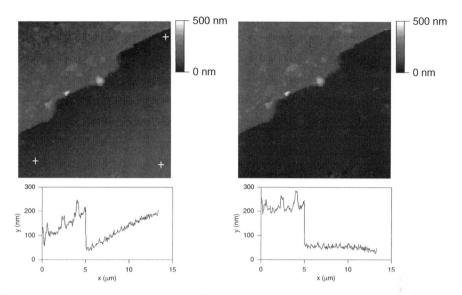

Fig. 5.2. Example of three-point plane levelling procedure. Left: result of plane-fitting to an image of an area of a sample with two distinct levels. The global plane fit does not give a good result (see the profile below), as it assumes the entire image is on the same plane. A global polynomial fit would not work well either. Right: result of three-point plane fitting using the three points indicated by crosses in the left image. The result is a much flatter background (see profile).

Exclusion of points from fit

Often AFM images show a few isolated features on a very flat surface. Typical examples include nanoparticles, micro-organisms on glass slides, or individual molecules on mica. With the most common type of levelling, polynomial line fitting, the presence of raised features on the substrate will cause a levelling artefact. This occurs because the algorithm fits the entire line, and sets it to the same level as all the other lines in the image. Thus, where large features occur on a line, the substrate becomes artificially lowered, and the image ends up with what look like 'shadows' or streaks behind any large features. Because these 'streaks' change the height of the background, they make feature height measurements inaccurate. This is quite a common artefact in processed AFM images, and many examples of images poorly processed in this way have been published. However, the problem is easily overcome by simply excluding certain regions from the fitting procedure. Typically this is done by a routine where the user draws boxes around the regions to exclude. In this way, only the real background will be used for the fitting, and correct levelling can be achieved as is shown in the Figure 5.3. The only problem with this technique is that it can be quite time-consuming. If there is more 'sample' than 'background' then it can be advantageous to use the software to select the areas to include in the analysis, rather than those to exclude. That is, the user draws boxes only over the substrate, rather than over the sample. In some cases (for example, a very dense sample of nanoparticles, or a sample with very irregular features), it can become extremely time-consuming and difficult to draw boxes around all the features in the image. In this case, some software has an option to automatically include or exclude features based on their heights,

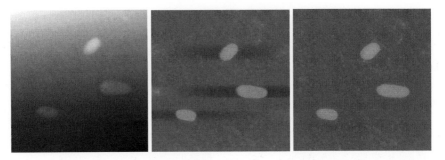

Fig. 5.3. Example showing how exclusion of features leads to improved polynomial line fitting-based levelling. Left: raw image of bacteria, with no levelling. Middle: the same image with second order polynomial levelling applied to the whole image – the procedure introduces artificial streaks around the cells, and can obscure surface features. Right: the image after second order polynomial levelling, excluding high features from the fitting – the background is completely flat.

for example in the case of particles on a flat surface, the upper 75% of the heights might be excluded, thus only the substrate would be used for levelling.

5.1.2 Filtering

Often there is unwanted high or low-frequency noise that appears in AFM images, and this noise may be removed by filtering. The two types of filtering most commonly used on AFM images are matrix filtering and Fourier filtering.

Matrix filters are based on averaging adjacent points in the image, such that certain frequency components are moved. Matrix filters are mostly grouped into so-called low-pass and high-pass filters. Low-pass filters are so-called because they allow low-frequency components of the image to pass, while reducing the high-frequency components. This has a smoothing effect on the data, and so removes the high-frequency noise commonly found in AFM images. Over-application of low-pass filters or use of large matrices will tend to blur the data. High-pass filters, on the other hand, allow high-frequency components to pass while reducing low-frequency components. They are usually used for sharpening or edge detection in AFM images. Examples of low- and high-pass filters are given below:

Filter name	Equation	Remarks
3 × 3 mean rectangular filter (the simplest smoothing filter).	$C = \frac{1}{9}\begin{bmatrix} 1 & 1 & 1 \\ 1 & 1 & 1 \\ 1 & 1 & 1 \end{bmatrix}$	4 × 4 or 5 × 5 matrices will give greater smoothing effects, and so on.
3 × 3 simple high-pass filter (also known as the unicrisp filter).	$C = \frac{1}{9}\begin{bmatrix} -1 & -1 & -1 \\ -1 & 9 & -1 \\ -1 & -1 & -1 \end{bmatrix}$	Larger matrices will tend to enhance the effect of these filters as well.

Examples of the results of applying these filters to an AFM image are shown in Figure 5.4. 5 × 5 matrices were applied, in order to enhance the strength of the effects on the image.

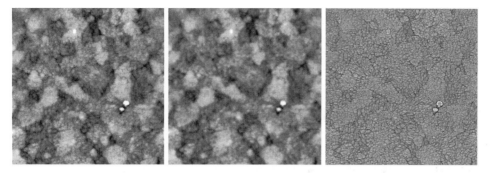

Fig. 5.4. Example of the results of matrix filtering. Left: unfiltered 2 μm image of ITO-coated glass. Middle: result of filtering the image with a 5 × 5 mean filter; this results in smoothing of the image and reduction of high-frequency noise and details. Right: result of filtering the image with a 5 × 5 unicrisp filter. This reduces low-frequency height differences, but greatly enhances edges and high-frequency features.

There are many other matrix filters that can be used, and are typically available in AFM processing software. These are more sophisticated versions of the filters discussed above, some of which, such as the median filter, can reduce noise without inducing edge-blurring as shown above for the mean filter [352]. Fourier transform-based filtering relies on transforming an image from real space into frequency space, removing certain components, and transforming back into an image in real space. Since Fourier transforms are used more often in AFM as an analysis tool than as a processing tool, they are discussed further in Section 5.3. In terms of their use as a filtering tool, Fourier techniques can achieve very similar results to matrix filtering, i.e. either high- or low-pass filters can be applied. However, Fourier filtering is somewhat more flexible, and it is possible to remove or to enhance specific components of an image, as shown in Section 5.3.4.

5.1.3   Rotation, cropping, and scaling

Rotation of images is a commonly available procedure in AFM analysis. This is necessary if it is required that features in an image line up with the scan axis. This can be required for the analysis of technical samples, for example for specific software analysis routines. In addition, occasionally it is necessary to scan a particular part of a sample before and after treatment. In this case, it can be useful to align the images before and after treatment.

A common treatment applied to AFM images is cropping, which could be used to remove unwanted features from the edges of the scan, or to isolate a particular part of the image for further analysis. For example, the roughness of different regions in a sample could be analysed by this technique. Pixelation caused by zooming into small regions of an image can make the image difficult to interpret. For display purposes, additional pixels can be added and a resampling algorithm function used to 'smooth' the image. However, this does not add any additional data to the image, and resampled images should not be used for further analysis.

Scaling in this context means changing the scale of an image. This is an operation that will rarely need to be used, as the calibration of the AFM instrument should be correct before images are recorded. However, under some circumstances it may be desirable to change the scale of an AFM image. For example, where an internal size standard in the image is present, and the measured dimensions are not accurate, this can be a way of correcting the calibration in the image. If images of the same feature with two different instruments are to be directly compared, it may be useful to rescale them so that they show features as exactly the same size. If the AFM is recalibrated during a set of experiments, it can also be useful to change the scale of the earlier images to match the known calibration.

5.1.4   Error correction

AFM images often contain 'errors', which typically result from unwanted interactions of the tip with the sample, such as sample movement under the tip, or strong tip–sample forces leading to vibration or streaking in the images. They can also be the result of external forces such as vibration or acoustic noise. If it's not possible to acquire a new image without the errors, the user may wish to remove them. This can be achieved by specific routines in the AFM processing software. Line removal can be done by removing a single line and replacing it with the average of the two lines next to it. A 'glitch' may be removed by replacing the glitched pixel with the average of the eight pixels around the unwanted pixel. These sorts of corrections should be used with great caution, and images corrected in this way should not be used for further analysis, as the correction process will change the data. For example, data that was subjected to error correction will exhibit lower roughness values after treatment, and these values do not reflect the true roughness of the surface. A further type of image correction commonly applied to AFM data is deconvolution. This process attempts to remove the effect of the finite width of the AFM tip (i.e. tip–sample convolution) from the AFM data. Almost all AFM processing software has routines to carry out deconvolution. This is best carried out when the shape of the tip is well known (either by blind estimation or measurement). Blind estimation of the tip shape is often also included in the software. The use of these procedures is described in greater detail in Section 2.5.

5.2   Displaying AFM images

A large amount of information can be gleaned from AFM images just by viewing them. However, the information visible in an image depends on how it is displayed. Furthermore, the AFM user never works in isolation, and needs to communicate the information present in the images to others. While much of this is done by the analysis techniques discussed in the next section, the ability to display images that show what we want is vital, to make use of AFM data. Often particular features can be enhanced and seen much more easily by changing display parameters. Although there are many different display functions in AFM data analysis software, in general the display functions don't alter the data in any way.

5.2.1   Histogram adjust

An image histogram is a plot of the height (or other parameter recorded in the z scale) of each pixel, versus frequency. The most common way of displaying AFM data is as an image where colour represents height. The AFM software will normally stretch the colour

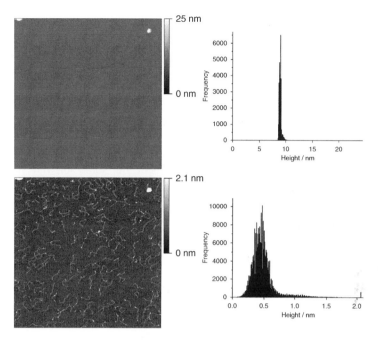

Fig. 5.5. Example of histogram adjustment. In the upper case, the data is mostly concentrated in the centre of the histogram; the image shows little detail. In the lower case, the colour range was adjusted so the data stretches over more of the available colour range; the features of interest (DNA molecules) are much clearer. The position of the y axis–0 nm height has also changed.

scale to cover the entire range of the z scale. If the user examines the histogram of height values, they often observe that 90% or more of the data is squashed into a narrow region of the histogram. This is often the case because of small amounts of outlying data points (i.e. very low or very high parts of the topography). These outlying regions may reflect real topographical features, but are sometimes caused by errors or glitches in the data. In either case, using a histogram adjust tool, the user may decide to reject upper and/or lower points from the colour scale, so that the colour scale is better distributed over the majority of the height data. This has the effect of greatly increasing the contrast in the majority of the image, and often helps greatly to visualize finer details in the image. An example of this use of the histogram adjust function is given in Figure 5.5.

   The height histogram can be useful for other functions apart from redistribution of image colours. For randomly varying topography, the histogram will usually display an approximately Gaussian distribution of heights. Deviations from the typical shape give information about the distribution of heights in the image. For samples with two significantly different regions, for example, there will be two peaks in the histogram. A staircase-like sample with various flat levels at different heights will give rise to further peaks. To selectively enhance the contrast on one of the two features, the user simply stretches the colour scale across the relevant peak in the histogram. The histogram tool can also be used for analysis–for example by measuring the distance between two peaks, the user obtains the average height difference between the two corresponding features. This analysis method should be used with caution, however. Although it has some advantages over other methods of measuring height differences between layers (see Section 5.3), it is very sensitive to the quality of the levelling, because poor levelling will tend to bring the peaks in the histograms closer together. In addition to histogram stretching functions, many AFM software packages contained brightness and contrast adjustment controls. These essentially perform the same functions as the histogram adjust function, but with somewhat less fine control over the result. Usually the AFM software will remember the whole resolution of the AFM image when performing histogram adjustment, but some software packages are set up to discard the extra data when a colour scale adjustment is made in this way, so the function should be used with caution if this is the case, as the data outside the new colour range could be lost.

5.2.2 Colour palettes

The colour palette used to display an AFM image can be selected to make the image seem more visually compelling. In some cases, selecting a specialized colour pallet can help with visualizing certain aspects of an image. For example, in the previous section, the special case of images with features of dramatically different heights was mentioned. While stretching the colour histogram to cover one set of features increases contrast on that feature, other features in the image lose contrast. By using a palette with more than one colour gradient, features at more than one height can be displayed while maintaining high contrast. This works because although the human eyes can distinguish less than 100 brightness levels of a particular colour, the combination of different shades allows many more to be distinguished [353]. An example of a case where a complex palette is useful is given in Figure 5.6. However, apart from the use of multiple colour gradients, the choice of the colour palette that is used for displaying an AFM image is very subjective.

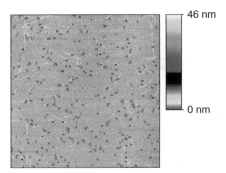

Fig. 5.6. Example of the usefulness of complicated colour palettes. In this image, the various shades allow features of different heights to be seen at the same time. Here, gold nanoparticles are seen in red and yellow, DNA in white, a salt layer in light blue and the substrate in dark blue. (A colour version of this illustration can be found in the plate section.)

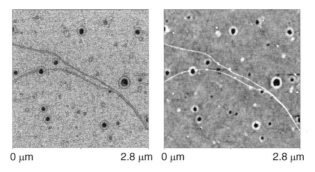

0 μm   2.8 μm 0 μm   2.8 μm

Fig. 5.7. Light shaded (left) and colour scale (right) images of carbon nanotubes. The light shaded image typically looks more photorealistic than the colour scale image, and helps the viewer to visualize the shape of features more easily.

5.2.3   Shading

One other option to enhance contrast and aid interpretation of AFM images is image shading. This routine applies an artificial shading filter to the data, and results in higher contrast, and a more photorealistic image, at the expense of height information. The position and intensity of the light shining on the AFM image can be changed. Light shading often helps visualize the smallest, high-frequency structure on a surface. The shape of image features is also often easier to appreciate in light-shaded images (Figure 5.7).

5.2.4   Three-dimensional views

AFM height data is inherently three dimensional (3-D). However, the standard method of rendering AFM data shows a two-dimensional (2-D) image, using a colour scale to represent height information. This is not a normal way for humans to see shapes, and can make interpretation difficult. In particular, for viewers unused to AFM data, it can be difficult to determine which features are higher than others, etc. One way to overcome this is to render the height information as a pseudo-three-dimensional image. This procedure is very quick and easy to implement with modern computers. Three-dimensional rendering has the effect of making the height information in the image simpler to understand. This is typically done in one of two ways; by maintaining the height colour scale, or including the effect of an imaged light source to illuminate the sample topography (similar to the shading discussed above for 2-D images). Maintaining the colour scale can make feature height interpretation simpler, while simulation of light source tends to highlight small features and texture on the sample topography. This latter method is somewhat more naturalistic. Finally, it is also possible to combine these two techniques, by height-colouring *and* lighting the topography. The choice of rendering, like choice of colours, is highly subjective, and 3-D images are very often used for 'artistic' or publicity purposes. However, in some cases such techniques really enhance the interpretation of the height data. Some examples of pseudo-three-dimensional renderings are given in Figure 5.8. In addition, it is also possible to produce true three-dimensional images with some packages. These typically require the viewer to wear

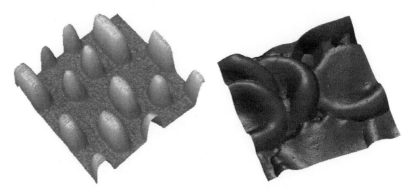

Fig. 5.8. Examples of 3-D renderings of height information. Left: DVD bits, right: human blood cells. Three-dimensional rendering greatly helps the viewer to visualize the shape of the sample, as well as to see both large and small features simultaneously.

special glasses to differentiate the left eyes and right eyes views, and can help to visualize the shape of the sample topography.

5.3   Analysing AFM images

AFM images are highly suitable for further analysis, and it is rare that only displaying the images acquired is sufficient to fully characterize the sample under study. Thus, very commonly further analysis is carried out in order to obtain quantitative information about the sample. Of course, the quantitative measurements derived from AFM images are only as good as the quality of the measurements. Thus, for the best data analysis the image must be acquired and processed properly. For example, if the images have errors due to poorly chosen acquisition parameters or dull probes, or have been inadequately processed the analysis will reflect these problems.

5.3.1   Line profiles

One of the most common analysis techniques is the extraction and measurement of line profiles. Because AFM images are difficult to measure dimensions directly from, line profiles are usually extracted in order to measure dimensions from the AFM images. The AFM analysis software allows the user to arbitrarily define lines to be extracted, and these can be horizontal, vertical or at any angle. The software then constructs a new plot, with distance along the chosen line on the x axis, and sample height on the y axis. In the new plot, the user can easily measure feature heights, widths, and angles. An example showing an image with an extracted line profile, and measurements of these parameters, is shown in Figure 5.9.

Figure 5.9 illustrates an important application of measurement of line profiles. Micro-fabricated devices such as these DVD master bits have very specific tolerances in their dimensions which directly affect their performance. Line profiles such as this can give very accurate measurements in such samples. It is important to be aware of the sources of error in such measurements. In an example such as this, the spacing between the bits

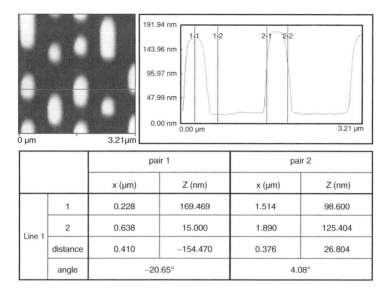

		pair 1		pair 2	
		x (µm)	Z (nm)	x (µm)	Z (nm)
	1	0.228	169.469	1.514	98.600
Line 1	2	0.638	15.000	1.890	125.404
	distance	0.410	−154.470	0.376	26.804
	angle	−20.65°		4.08°	

Fig. 5.9. Line profile analysis of an AFM image of DVD bits. The height (indicated by Z in the image) and width (indicated by X) of the surface features can be measured from the line profiles.

should be accurate, as is the height of the features. However, the width of the features is subject to convolution with the tip shape as described in Chapter 2, and is thus likely to be somewhat larger in the line profile than in reality. Deconvolution (see Section 2.5) or using speciality tips can help to improve the accuracy of these measurements [53, 354, 355]. Feature height in AFM is generally the most accurate measurement (especially for relatively incompressible samples such as the DVD), and thus height is the dimension most commonly measured in line profile analysis. Like all analysis methods, the measurement of feature heights is sensitive to the prior image processing. In particular, levelling should be done as accurately as possible (avoiding levelling artefacts, see Section 6.3.1) before line profiles are measured. Line levelling artefacts can artificially lower large features on the surface, and thus can lead to underestimation of feature height. If a horizontal polynomial fitting routine was used, then measuring the profiles horizontally can reduce the effects of this problem.

Manual measurement of line profiles is a very simple and accurate way of measuring features in AFM images. However, if a large number of features are to be measured, it is rather laborious and time-consuming. The results can also be somewhat operator-dependent, as the exact spots on the profiles that the operator chooses to define the features of interest can vary somewhat from user to user (see Figure 5.10). For the measurement of large numbers of features, more automated routines (such as particle analysis, discussed below) can be more efficient. In order to remove operator dependence, some well-defined algorithms exist to measure, for example step heights, or feature heights and widths [356, 357]. An illustration of a commonly used ISO-defined algorithm to measure step height is shown in Figure 5.10.

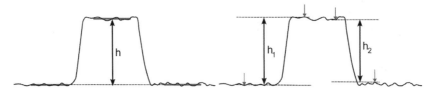

Fig. 5.10. Feature height measurement techniques. Left: ISO 5436 profile measurement [357]. Thick lines show measured regions; the rest of the data is discarded to avoid influence from edges. The measured parameter is h. Right: Illustration of the problem with simple line-profile measurement. The small arrows indicate points in the line profile that the operator might choose in order to measure the height. The selection of these points can affect the results: $h_1 \neq h_2$.

5.3.2 Roughness

Roughness is a very important surface property for technological applications, but is also widely used as a quantitative measurement of surfaces changes in a wide range of samples. For example, absorption of materials to a surface, or erosion of a surface will typically lead to an increase in roughness. There are a number of different ways to measure roughness, and it can be measured on lines or in particular regions within images, though whole images are most usually measured.

The most commonly used roughness parameters are probably the arithmetic roughness, (R_a) and the root-mean-squared roughness (R_q or R_{rms}). Both of these values have a positive correlation, i.e. larger values mean greater topographical variation in the image. Often, both values give rather similar results, but R_q will always be somewhat larger than R_a, and is rather more sensitive to outlying points than R_a. The formulas for these parameters along with some other commonly calculated roughness parameters are shown in Table 5.2. Roughness, measured by R_a or R_q, is one of the most commonly used statistical parameters for describing AFM images. It gives useful information about the sample surface, and can be correlated with results from other techniques [358–360]. Therefore the value of these roughness parameters can be considered characteristics of the

Table 5.2. Some commonly used roughness parameters [363].

Parameter name	Abbreviation	Formula
Roughness average	R_a	$R_a = \frac{1}{n}\sum_{i=1}^{n}|y_i|$
Root-mean-squared roughness	R_q or R_{rms}	$R_q = \sqrt{\frac{1}{n}\sum_{i=1}^{n}y_i^2}$
Maximum height of the image (peak to peak height)	R_t	$R_t = \left|\min_{1\leq i\leq n} y_i\right| + \left|\max_{1\leq i\leq n} y_i\right|$
Skewness	R_{sk}	$R_{sk} = \frac{1}{nR_q^4}\sum_{i=1}^{n}y_i^4$
Kurtosis	R_{ku}	$R_{ku} = \frac{1}{nR_q^4}\sum_{i=1}^{n}y_i^4$

sample, not just of the technique. However, image processing history is important. The degree of levelling applied to an image directly affects the roughness values obtained [361], as does image size [359, 362], so all data should be obtained and processed identically for comparison of roughness values to be valid. Some examples and more details of roughness measurements using AFM images are given in Section 7.1.1.

While R_a, R_q and R_t all describe the magnitude of the roughness–larger values mean rougher surfaces–skewness and kurtosis describe the distribution of the sample height data. Skewness is a measure of the asymmetry of the distribution of heights. Random variations in topography will give rise to a skewness value of 0. Positive values of skewness indicate the presence of height values considerably above the average, while negative values indicate the presence of height values considerably below the average. Thus a flat surface with protruding features would have a positive skewness, while pits or depressions in the surface would lead to negative skewness. Kurtosis, describes the 'peakedness' of the distribution of height values. A distribution with high kurtosis would have a small number of extreme heights (i.e. a few very high peaks or very low valleys), as opposed to many moderate height features (which will give lower, or negative kurtosis values). A Gaussian or normal distribution has a kurtosis value of 0. Skew and kurtosis are used with AFM data much less commonly than the R_a and R_q parameters, but are sometimes useful in characterization of surfaces [364–366].

5.3.3 Particle and grain analysis

Nanoparticle analysis is one of the most popular applications for AFM, due to the ability to measure accurate particle dimensions with sub-nanometre accuracy. In order to generate adequate statistics to characterize a sample of particles by any microscopy technique, it is necessary to measure a large number of particles, typically in the hundreds, or even more. For a small number of particles, the height and width of the particles can be quickly measured with the line profile tool. However, for large samples, this rapidly becomes tedious. For this reason, most AFM analysis packages include routines for automatic particle counting and analysis. As explained in Section 5.1.1, images with particles on a flat background need care to be levelled properly. This is particularly important if the image is to be analysed by an automated routine. However, once good levelling is achieved, identification, counting, and measuring can be performed with an automatic particle analysis routine. For spherical particles on a flat, well-levelled background, a simple thresholding routine can isolate the particles. In this process, the user selects a value above which all features are counted as particles. The process of separating the particles from the background is known as image segmentation. In addition to the comparatively simple thresholding routine, there are a number of more sophisticated segmentation routines which can be applied in non-ideal cases [367]. The reason a large number of segmentation routines exist is that simple threshold segmentation is not a very robust method for non-ideal cases. Another common application of such routines is identification of grains in a sample surface. Because granular materials do not usually have great height differences at the grain interphases, height thresholding does not usually work very well for these samples. Other routines that are based on changes in slope work rather well for grain analysis, as does the so-called 'watershed' segmentation method [368, 369]. This latter routine simulates water drops being placed in the image, and then

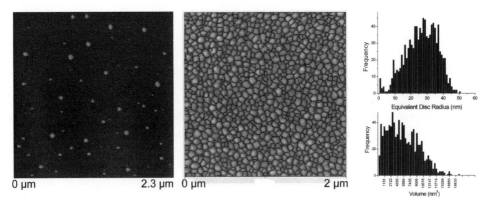

Fig. 5.11. Examples of (left) particle counting and (middle) grain analysis; in both cases the automatically detected features are highlighted in red. On the right are shown some typical results – histograms of grain volume and the radius of an equivalent disk. (A colour version of this illustration can be found in the plate section.)

flooding until they meet. This typically gives far superior results in segmenting grain images.

However it is carried out, once the segmentation is complete the AFM analysis software measures particle or grain heights, widths, radius, areas and even volumes. Usually it is possible to export all these measurements for further statistical analysis. The software typically does not count 'partial' particles that are clipped at the edge of an image. Some packages also allow direct calculation of the statistics (e.g. mean average, standard deviation, etc.) and plotting of histograms of these parameters. Usually, the user would like to combine the results from several images together to improve the number of samples, so it is most convenient to export the data for consolidation and further analysis, e.g. in a spreadsheet. Particle analysis routines can also be applied to measurements of pits or depressions in a surface by simply inverting the image during the processing step. An example of a grain/particle counting routine and the results are shown in Figure 5.11.

5.3.4   Fourier transform and autocorrelation analysis

In AFM image processing and analysis, a two-dimensional Fourier transform is an operation that converts the AFM image from the spatial domain, into the frequency, or more correctly, the wavelength domain. This is carried out by a mathematical operation known as a fast Fourier transform, so is sometimes also known as FFT analysis. When transformed into Fourier space, the image will show features in terms of wavelength (or frequency). This is particularly useful to identify any repeating patterns in the image. For example, the Fourier transform may be used to identify the frequency of noise in an image. Most unwanted sources of noise have a characteristic frequency; once the frequency is identified via FFT analysis, it is easier to identify the source of the noise and eliminate or suppress it. The second major use of the Fourier transform in image analysis is to extract useful data about the sample. This has been used for roughness analysis [370] and to identify frequency characteristics of large features [371], but is probably most commonly

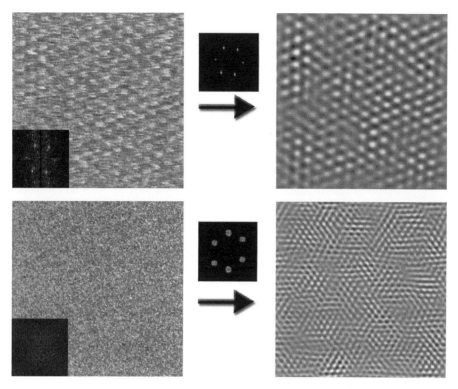

Fig. 5.12. Examples of use and abuse of Fourier analysis and filtering. Top: the noisy image of an atomic lattice on the left shows clear spots in the FFT image (inset). Measuring these spots allows the user to determine the lattice parameters. Filtering is achieved by selecting the six intense spots, as shown above the arrow, and removing the rest of the image in Fourier space. Carrying out an inverse FFT gives the image on the right – the selected features are preserved and the noise is removed. Unfortunately, as shown below, carrying out the same operation on an image of pure noise (left, note the inset FFT image shows no spots) can also give a wholly artificial image (right). (A colour version of this illustration can be found in the plate section.)

used to analyse atomic lattice parameters [372, 373]. Images showing the atomic lattice, even if very noisy, will tend to show intense spots in the Fourier space image, indicating the unit cells of the atomic lattice. An example of this is given in Figure 5.12.

In order to determine the spatial wavelength of the features seen in the Fourier-space image, they can be measured directly with the analysis software, or an alternative is to use Fourier filtering to isolate the components of interest. The way this is done is by editing the image in Fourier space directly. Typically, the user draws boxes (or ellipses) in the Fourier image around the features of interest. The software keeps these marked areas, and clears the rest of the image. Then an inverse Fourier transform is applied to the modified image. This has the effect of converting the image back into the spatial domain, i.e. into real space, but with only the selected components remaining. This removes all other features from the images, leaving only the features the user selected, allowing clear visualization or measurement of the pattern of repeating features. As may be imagined, this technique is

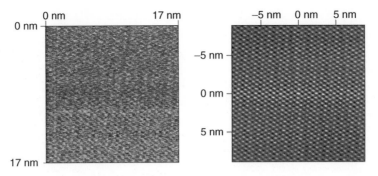

Fig. 5.13. Example of autocorrelation analysis. Left: unfiltered noisy image of an atomic lattice. Right: 2-D autocorrelation function derived from the image. Note the change in scale; the origin moves to the centre of the image, because the result is not an image, but rather it's a plot of self-correlation in the source image. Typically, the centre will be clearest, showing short-range correlation; long-range correlation is required for the outer regions to have the same amplitude.

also a very powerful filtering method. For example, one might be able to use this method to remove noise at any particular frequency from an image. Other examples include removal of streaking in images of soft samples [374]. However, one must be very careful in interpretation of Fourier-filtered images; it is easy using to such a method to artificially create an image with any desired pattern, even one that never existed in the original un-filtered data; see the lower part of Figure 5.12.

Autocorrelation analysis is another method for the analysis of repeating patterns in data. Outside of AFM, one-dimensional autocorrelation is most popular, but for AFM images, two-dimensional autocorrelation is more useful. Two-dimensional autocorrelation of an image results in an image containing the periodic patterns present in the original image, with spontaneous (non-repeating) features removed. Unlike the Fourier transform image, the autocorrelation function is shown in real space, so the spacing of the repeating features can then be measured directly from the image. Like FFT analysis, a common application for autocorrelation analysis in AFM is measurement of atomic lattice spacing. An example of two-dimensional autocorrelation analysis is shown in Figure 5.13.

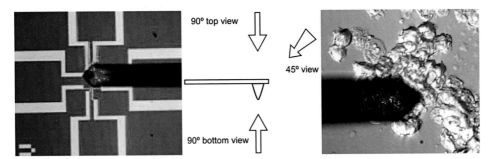

Plate 1. Left: video camera image of the cantilever and sample in an AFM (90° top view). The red 'spot' is from the laser that is used in the optical lever force sensor. With scanning ranges greater than 1 μm, it is possible to see the AFM cantilever move in the video microscope image. Middle: the three possible viewing positions of an optical microscope in an AFM. Right: image in an AFM with 90° bottom view; note the laser light (purple in this case) can be seen through the cantilever, which is seen through the sample (cells on a glass slide).

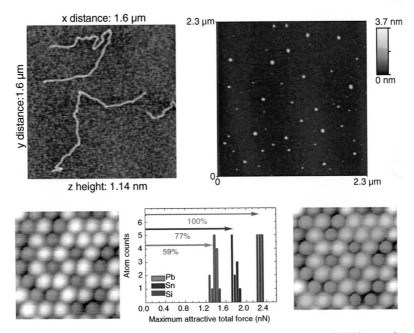

Plate 2. Example non-contact AFM images. Top: examples of non-contact AFM images in ambient conditions (air) – individual DNA molecules (left) and 1 nm nanoparticles (right) [123]. Bottom images: non-contact AFM in UHV conditions for individual atom identification. Left: atomically resolved NC-AFM image of Si, Sn and Pb atoms on an Si(111) substrate – some atoms may be differentiated based on apparent size, but identification is not possible. Middle: short-range chemical force measured over each atom is dependent on the chemical nature of the atoms. Right: the same image as on the left, with atoms coloured according to the colour scheme in the middle. Adapted from [8], with permission.

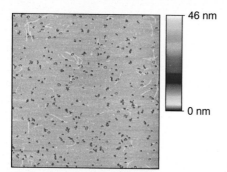

46 nm

0 nm

Plate 3. Example of the usefulness of complicated colour palettes. In this image, the various shades allow features of different heights to be seen at the same time. Here, gold nanoparticles are seen in red and yellow, DNA in white, a salt layer in light blue and the substrate in dark blue.

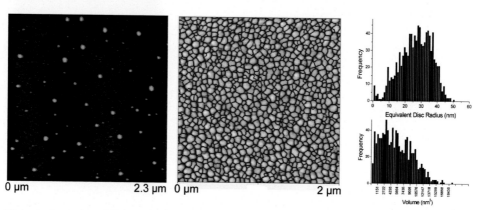

0 μm   2.3 μm   0 μm   2 μm

Plate 4. Examples of (left) particle counting and (middle) grain analysis; in both cases the automatically detected features are highlighted in red. On the right are shown some typical results – histograms of grain volume and the radius of an equivalent disk.

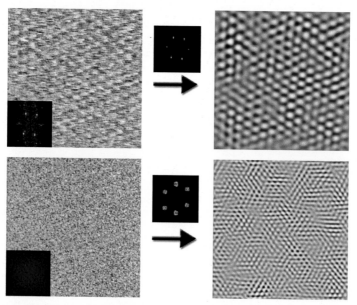

Plate 5. Examples of use and abuse of Fourier analysis and filtering. Top: the noisy image of an atomic lattice on the left shows clear spots in the FFT image (inset). Measuring these spots allows the user to determine the lattice parameters. Filtering is achieved by selecting the six intense spots, as shown above the arrow, and removing the rest of the image in Fourier space. Carrying out an inverse FFT gives the image on the right – the selected features are preserved and the noise is removed. Unfortunately, as shown below, carrying out the same operation on an image of pure noise (left, note the inset FFT image shows no spots) can also give a wholly artificial image (right).

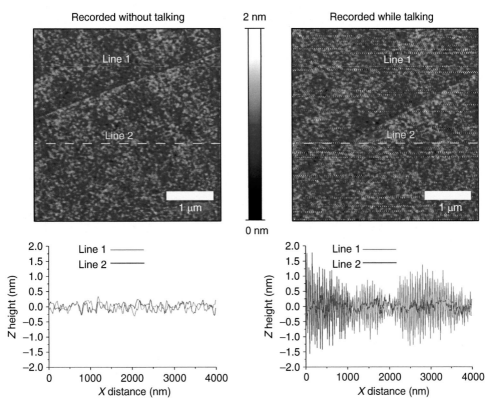

Plate 6. Effect of acoustic noise. This high-resolution image of a test grid shows the effect of acoustic noise on an image. Right: image and line profiles measured while acoustic noise was present in the room. The acoustic vibrations from a person speaking while the image was acquired are clearly visible in the line scans and the image. Left: image that was measured without the acoustic noise.

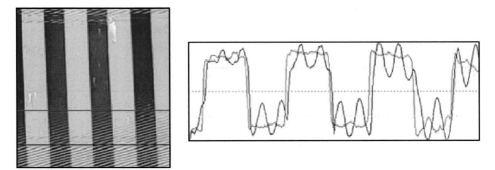

Plate 7. Example of electronic noise in an AFM image. This image of a test pattern has electronic noise at the top and bottom of the scan. The electronic noise in this case was a result of not having a ground wire attached to the stage. The artefact was identified by the oscillation frequency.

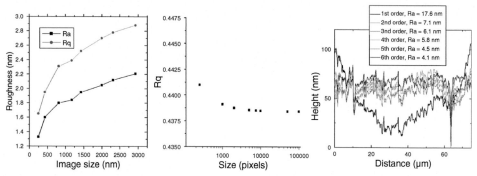

Plate 8. Illustrations of the effects of scanning and data processing parameters on measured surface roughness. Left: the effect of changing the area scanned (image size) on the measured roughness values (R_a and R_q). In general smaller AFM scans show smaller values of roughness. Centre: the effect of changing the number of pixels in the image (pixel resolution) on the roughness (R_q). In general, there's a very weak relation between the pixel resolution and the roughness value. Reproduced with permission from [415]. Right: effect of image processing (levelling algorithm) on roughness of line scans on a femoral head implant replica (R_a). The higher the order of polynomial applied, the lower the roughness [361]. Reproduced with the kind permission of Dr. James R. Smith, University of Portsmouth, UK.

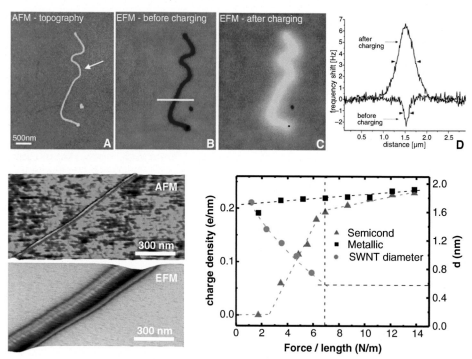

Plate 9. Electrical measurements of carbon nanotubes. Top: charge injection into nanotubes. **A**: topography of isolated MWNT, and its EFM image before charging (**B**). After charge injection by the AFM probe (at the arrowed location), the EFM contrast reverses(**C**). **D**: a line profile through the EFM images at the point indicated by the white line. Bottom: measurement of the effect of mechanical compression on charging behaviour of SWNTs. Left: AFM and EFM (measured in lift mode) images of a SWNT after charge injection by the AFM tip. Right: plot showing charge density λ (red triangles) as a function of compression of the semiconducting CNT by the probe (indicated by the diameter, d, which decreases as the CNT is compressed, green circles). The black squares show the behaviour of a metallic SWNT, illustrating how compression with the AFM changes the behaviour of the semiconducting SWNT to match the metallic behaviour. Adapted with permission from [550] and [558]. Lower figures copyright (2008) by the American Physical Society.

Chapter 6

AFM image artefacts

All measurements and measurement techniques are prone to artefacts. In AFM imaging, these artefacts are sometimes easy to spot and sometimes very difficult. Some artefacts can be easily avoided, if the user knows what to look for and knows the source of the error. A few artefacts are unavoidable, but knowing that they exist in an image helps to avoid misinterpreting them as genuine image features. This means that recognizing image artefacts is very important for the AFM user. However, when users begin to use AFM for the first time, it is very difficult to sort the real features from the artefactual. Experienced AFM users as well as novices can benefit from considering the sources of AFM artefacts, as some artefactual features are very subtle, and can only be clearly seen when making particular measurements from an image (for example when measuring line profiles, or Fourier filtering). This chapter contains examples of common AFM artefacts, explains the source of the features, and shows what can be done to avoid them.

6.1 Probe artefacts

Probably the most commonly seen AFM artefacts arise from the probe used to scan the sample. As explained in Chapter 2, all AFM images are a convolution of the topography of the sample with the shape of the tip of the probe (and sometimes with the sides of the probe) [54]. When interpreting AFM images, we often assume that the tip radius is finer than the details imaged, and that the opening angle of the probe is smaller than the angle of the features in the sample. This means that the influence of the tip-shape on the image obtained will be small (but finite). However, even if this is the case, continual use may dull the probe tip or it can break or become contaminated [46]. Often, if the user has many samples to image, the probe will be used until one of these phenomena occurs, and the probe becomes unusable. In either case, the user must know what to look for when the tip degrades, in order to know when to replace the probe.

Common effects seen when imaging with an inadequate probe include:

- The features on a surface appear too large.
- The features, especially holes, appear too small.
- Strangely shaped objects appear.
- Repeating patterns appear in the image.
- The image appears normal on the top of features, but not on their sides.

The best advice if the user is unsure is to use a tip-check sample. This can be any sample that the user is certain of the topography of, and which has relatively fine features, such that the radius of the tip can be determined. In practice, certain types of samples are particularly useful for this operation, and some of the most common ones are described in

Appendix A. In this chapter, images of tip-check samples that were acquired with faulty probes are shown, along with images measured with a new probe, to illustrate the effect that probe damage has on the images obtained.

6.1.1 Blunt probes

Typically, blunt probes will lead to images with features larger than expected, with a flattened profile, due to the effect shown below. Note that holes in a flat surface will show the opposite effect, appearing smaller with blunt probes than with sharp ones (see the lower part of Figure 6.1).

The dilation due to the probe shape as shown in Figure 6.1 is a normal feature of AFM imaging. For example, when measuring globular features with a known diameter of 2 nm it would be normal to find the feature in the AFM image has 2 nm height but 10–20 nm width [279, 375, 376]. However, when it occurs to a large extent it is a problem, because it may significantly alter the apparent size of the features, and can really change their appearance. An example is shown in Figure 6.2. If this effect is noticed, the user should change the probe. If the feature cannot be imaged correctly even with newer probes, then another type of probe (e.g. super-sharp probes or high-aspect-ratio probes) may be required [377]. However, some extremely high-aspect-ratio features can be extremely challenging to image by AFM, no matter which probe is chosen.

The fine details of the BOPP sample when imaged with a sharp probe are seen in the left image in Figure 6.2. When imaged with a blunt, worn probe, as shown in the right image, they disappear, and the sample becomes almost unrecognizable. An example of the effect of pits in a sample becoming smaller with a dull probe is shown in Figure 6.3.

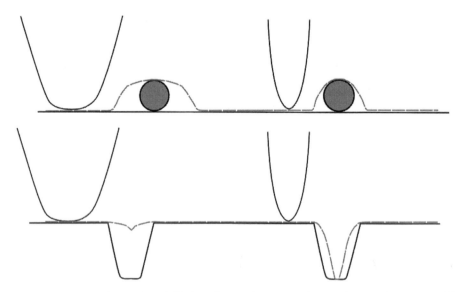

Fig. 6.1. Illustration of probe-based dilation. Convex features such as particles tend to appear wider with blunter probes, although feature height may be accurate. Concave features such as pits tend to appear smaller (both less wide and less deep) with blunter probes.

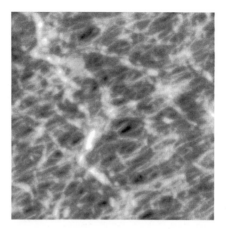 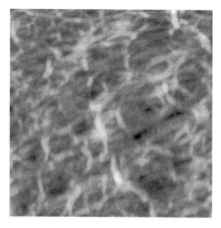

Fig. 6.2. Illustration of the effect of using a blunt probe. These two images are of the same sample, and both are 1.5 μm × 1.5 μm × 40 nm. The image on the left was taken with a sharp probe, the image on the right with a blunt probe. The sample is BOPP, a useful sample to characterize the sharpness of IC-AFM probe tips, see Appendix A9.

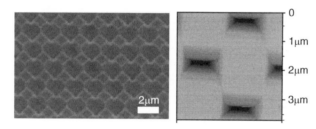

Fig. 6.3. Example of features appearing smaller due to the use of a blunt probe. Left: SEM image of a test pattern of squares (NT-MDT grating TGX, see Appendix A). The sides of the squares are all equal. B: AFM image of the test pattern. Because the probe is not sharp, the test pattern squares appear much smaller than they should, and appear as rectangles instead of squares.

6.1.2 Contaminated or broken probes

Contamination of AFM probes is quite common, and scanning certain samples leads to dirty probes more quickly than others. In particular, biological or other soft samples, or any sample with loose material at the surface, tend to contaminate probe tips quickly, leading to image degradation [378]. On the other hand, breaking of the AFM probe is less common, but still occurs, mainly when the probe accidentally touches the sample outside of feedback control. The reason these two problems are described together is than they can give very similar results. When imaging a sample with a broken or dirty probe, the resulting images often contain features with unexpected shapes, due to convolution of the misshapen tip with the sample features. Examples are shown in Figure 6.4. Any repeating patterns within the images, which are not expected based on what is known of the sample, are likely to be due to a broken or contaminated probe.

Fig. 6.4. Examples of how images produced with broken or dirty probes show repeating patterns in the images. Left: SEM images of damaged and dirty probes. Right: AFM images produced using the probes shown on the left. Images with repeating patterns like these are usually due to broken or dirty probes.

Double tips

A further example of damage or contamination of tips altering the image is the creation of multiple tips. If the tip breaks such that it has small spikes at the end, or more commonly, has debris attached near the tip, the sample may be imaged both by the true tip, and the debris. This results in multiple copies of each feature appearing in the image [379]. It's not possible to distinguish which image feature is from the 'true' tip, and double, or multiple copies of each feature occur in the image, as shown in Figure 6.5.

When the user determines that the probe is blunt, contaminated, or broken, they must replace the probe. Some procedures for cleaning of AFM probes have been described [380], however, in the authors' experience, it is usually simpler and far more effective to replace the probe than to try to clean it.

Fig. 6.5. Example of double-tip imaging. Left: an image of vesicles measured with a dirty tip. Right: DNA molecules measured with a broken tip, each molecule has a false 'twin' next to it. Centre: a badly broken and contaminated tip which produced double-tip images like these.

6.1.3   Probe–sample angle

When scanning large features, artefacts can be introduced by having a large angle between the probe and the sample, as illustrated in Figure 6.6. Ideally, the AFM probe should be perpendicular to the sample surface.

Solving this problem is achieved by adjusting the angle between the probe and the sample so that they are perpendicular. Often, a set of three adjustment screws on the microscope allows the user to adjust this angle. In many microscopes the probe is designed to be at a $12°$ angle with respect to the sample, and some probes are designed with this angle in mind, i.e. such that when the cantilever substrate is at $12°$ to the sample, the probe will be perpendicular to it. Some AFMs do not have mechanical adjustments to control the probe–sample angle. In this case, the sample must be adjusted to correct the probe–sample angle.

6.1.4   Side-wall/probe imaging

Certain samples with extremely high-aspect-ratio features are very difficult to image correctly, and they can interact with the probe in such a way that the image contains repeating images of the probe, or of the side-walls of the probe. Examples of features that produce side-wall images are spherical micro-organisms, spherical particles or red blood cells, with their typical doughnut-like shape, images of which often are great on the top of the cell, but it's not possible to image the sides of the cell, and images of the probe side-

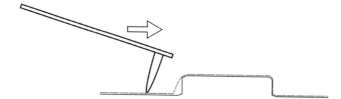

Fig. 6.6. Illustration of probe–sample angle problems. With the probe at an angle to the sample, distortions are introduced, and sample features appear asymmetric.

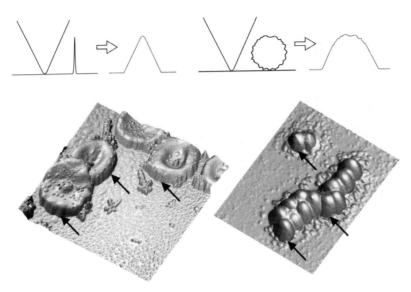

Fig. 6.7. Effect of probe or probe side-wall imaging. Top: illustrations of the effect of imaging a spike – an image of the probe is produced – and imaging a sphere-like feature – only at the top is the sample topography reproduced, and the rest of the image feature shows the probe's side-walls. Bottom: examples of probe–side-wall imaging. Left: red blood cells, right: *S. aureus* bacteria. In both cases, only the upper parts of the cells can be imaged correctly (some examples of probe side-wall imaging highlighted by arrows).

wall appear instead (see Figure 6.7) [381, 382]. Samples with spike-like features (including certain tip-check samples, see Appendix A) lead to repeated copies of the tip in the resulting images [383].

Typically, any image showing square pyramid-shaped features will be showing images of the probe rather than true sample features, so these image features can be discounted. In order to avoid this problem, the user is recommended to use a shaper tip, specifically, one with a higher aspect ratio. Silicon nitride contact-mode probes are very prone to producing images of the probe side-wall, as they typically have much wider opening angles (*ca.* 35–40° versus 15–20° for most intermittent contact-mode probes). If this artefact causes real problems, for example in metrology applications, super-high-aspect-ratio probes are also available (for example, with opening angles <3°) [377]. Example images of such probes are shown in Figure 2.30. However, for spherical samples such as nanoparticles or the cocci shown in Figure 6.7, parts of the sample will always be unavailable to most AFM experiments. Imaging of probe side-walls will tend to increase if there is a mismatch between the angle of the probe and the sample, as described in the previous section.

6.2 Scanner artefacts

As described in Chapter 2, there are a number of different scanner designs available for commercial AFMs. However, by far the most common design in use is the piezoelectric tube scanner. This scanner is used because it is easy to integrate into the instrument, cheap

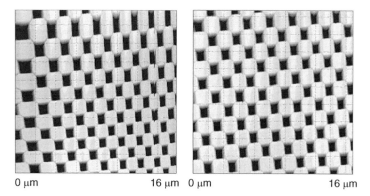

0 µm   16 µm 0 µm   16 µm

Fig. 6.8. Effect of *x-y* non-linearity in AFM images. Left: example of an AFM image of a test sample (TGX01, see Appendix A) when scanned without correctly linearizing the AFM scanner. Right: linearized AFM image of the same sample. The spacing of the squares at the top, bottom, left and right sides should be all the same distance apart. Images courtesy of Mikromasch.

to produce, and gives rapid and very precise response under most circumstances. However, most AFM scanners do introduce some artefacts into the images obtained, the tube scanner more than most. The artefacts described in this section all occur with piezoelectric tube scanners. Many of them are avoided when using a linearized scanner (see Chapter 2).

6.2.1 X-Y *calibration/linearity*

All atomic force microscopes must be calibrated in the *X-Y* axis so that the images and measurements obtained are accurate. The motion of the scanners should also be linear so that the distances measured from the images are accurate. Due to the non-linearity of piezoelectric scanners, without correction, the features on an image will typically appear smaller on one side of the image than on the other, see Figure 6.8. Once the scanner is properly linearized, it is also critical that the scanner be calibrated. In other words, it is possible for the scanner to be linear but not calibrated. If the calibration is incorrect, then the *X-Y* values measured from line profiles will be incorrect.

A common method for correcting the problems of *X-Y* non-linearity and calibration is to add calibration sensors to the *X-Y* piezoelectric scanners. These sensors can be used to correct the linearity and the calibration in real time; often, such a system is described as having linearized scanners. If these are not available, and non-linearity is detected in images, then the instrument should be re-linearized according to the manufacturer's instructions. Typically this is carried out with a test grid as illustrated above, and in Appendix A. Note that non-linearity at just one edge of the image could be due to other effects; see the other sections in this chapter.

6.2.2 z *calibration and linearity*

Height measurements in an AFM require that the piezoelectric ceramics in the Z axis of the microscope are also both linear and calibrated. Usually the microscope is calibrated at only

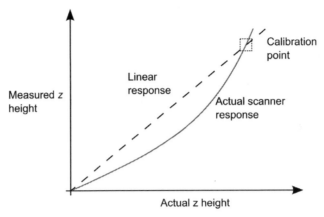

Fig. 6.9. Graph showing the relationship between an actual z height and a measured z height in an AFM. Usually only one calibration point is measured as shown by the box, and the z piezoelectric is assumed to be linear, as shown by the dashed line. However, as is often the case, the piezoelectric actuator is non-linear, as shown by the solid line. In such cases incorrect z heights are measured unless the feature being measured has dimensions close to those of the calibration specimen.

one height. However, if the relationship between the measured z height and the actual z height is not linear, then the height measurements will not be correct, see Figure 6.9.

The only way to ensure absolutely accurate z height measurements at a range of heights is to use an instrument with a sensor for the z piezoelectric. An alternative, which only works for measurements of features within a particular height range, is to recalibrate the instrument using a calibration specimen of known height, which is similar in size to the features which will be measured. Typically the z axes of AFM microscopes are calibrated using semiconductor test samples with features on the order of 100–200 nm in height. So, for example, measurements of small features of 5–10 nm could not be expected to be very accurately measured under these circumstances. In this case, it would be best to recalibrate the instrument using a test sample of known height in the range 5–10 nm. Alternatively, some samples can be used as an internal standard, avoiding the need to recalibrate the AFM [279]. Some widely available Z-height calibration standards are described in Appendix A.

6.2.3 Scanner bow

The scanners used in AFM instruments often move the probe in a slightly curved motion over the sample surface. This is typically the case for tube scanners fixed to the microscope body at one end, and free to move at the other – currently the most common design in AFM. As shown in Figure 6.10, this motion gives rise to a curvature or 'bow' as it is most often known, in the resulting images. This tends to give a small variation in z height over a relatively large X-Y area, so it is most obvious with flat samples.

This artefact cannot be avoided with instruments whose design is prone to it, but the effect can be removed in processing. The procedures to carry out this operation are described in Section 5.1.1.

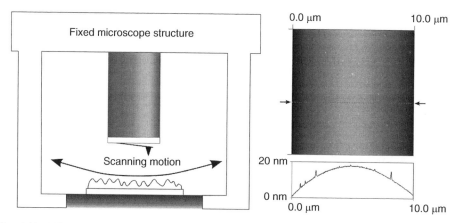

Fig. 6.10. Effect of scanner bow. Left: with tube scanners fixed at one end, the trajectory of the probe is curved. Right: the result is an apparent curvature in the height of measured samples, although the height change is small over a large area.

6.2.4 Edge overshoot in the Z axis

Hysteresis in the piezoelectric ceramic that moves the cantilever in the perpendicular motion to the surface can cause edge overshoot. Hysteresis is an inherent property of piezoelectric materials, and means that forward and backward movements are not exactly equivalent. The effect in the Z axis affects the AFM's ability to trace accurately over step profiles. This problem is most often observed when imaging microfabricated structures such as patterned Si wafers or compact disks, but may be observed in any sample with sharp-edged features. The effect can cause the images to appear visually better because the edges appear sharper. However, a line profile of the image structure shows errors, as shown in Figure 6.11.

Edge overshoot cannot be avoided by the user. It will only occur on microscopes without a z axis calibration sensor, however. In cases where this occurs step height measurements should only use the unaffected (flat) portion of the feature profile.

6.2.5 Scanner creep

Creep in piezoelectrics gives rise to the phenomenon that when an instantaneous voltage is applied to the piezoelectric and maintained, the response of the material does not follow exactly the applied voltage, but instead continues to move in the same direction as the initial offset, even when the voltage is no longer changing. This is illustrated in Figure 6.12. The practical effect of this is that when the user translates the scanning position on the sample, moves the probe to the start of a new scan, or zooms into a previous scan (all of which are done by rapidly changing the voltage applied to the piezoelectric), distortion occurs in the image. The duration of this effect is limited, and eventually it disappears. An example of this distortion ('scanner drift') is shown in Figure 6.12.

This artefact can be removed by simply waiting for the piezo position to stabilize. One way is to make an initial scan in any new region, before recording a second scan free of

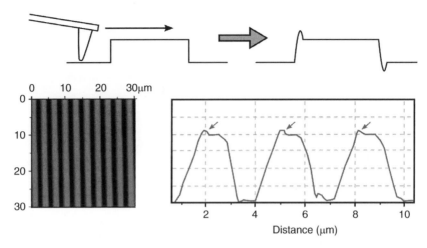

Fig. 6.11. Edge overshoot in the z axis. Top: the probe is scanned from left to right across a feature on a surface; overshoot may be observed in the line profile at the leading and trailing edge of the features. Bottom: the AFM image of a test pattern appears to have no artefacts at first glance (left), but a line profile of the test pattern shows overshoot at the top of each of the lines (right, overshoot arrowed).

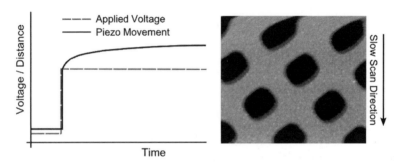

Fig. 6.12. Scanner drift cause and effect. Left: creep in piezoelectric scanners causes the scanner to keep moving even after the applied voltage stops changing. Right: the effect on AFM images is most often seen as a distortion in the beginning of the scan (here, scanning from the top).

distortion. Alternatively, the instrument can be set to do a continuous line scan in the new position. When the user observes that the features in the line scan are no longer changing, the drift has stopped and the image scan should then be begun.

6.2.6   Z angle measurements

Mechanical coupling between the piezoelectric ceramics that move the probe in the x or Y directions and the Z direction can cause substantial errors when trying to measure vertical angles with the AFM. This sort of crosstalk is common in piezoelectric tube scanners, and means that the accuracy of angles in the Z axis measured with most AFMs

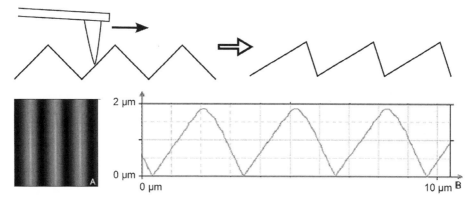

Fig. 6.13. Illustration and example of errors in Z angle measurement by AFM caused by crosstalk. Top: illustration of the effect. The sample has a series of repeating triangles at its surface. A line profile of the sample shows that the triangles do not appear symmetric. Bottom: real AFM image of a sample having a triangle pattern at its surface, and a line profile extracted from the AFM image. Although the angles of the two facets are in reality equal, the AFM image suggests that this is not so.

are unreliable. This error can best be measured with a sample that has repeating triangle structures. An example of this is shown in Figure 6.13.

The user cannot control the appearance of this artefact. It occurs with non-linearized tube scanner-based AFMs, and independent X-Y and Z scanners are required for the measurement of correct Z angles.

6.3   Image processing artefacts

Some image processing is usually necessary before viewing or analysing any AFM image. As described in Chapter 5, there are a large number of processing operations that can be applied to AFM images. The correct procedures were described in Chapter 5, so here only examples of the artefacts that might be introduced are shown.

6.3.1   Levelling artefacts

Levelling changes the entire AFM image, so the resulting image is different from the raw data. However, it is very often a necessary procedure before useful information can be extracted from an image. Commonly, levelling artefacts are introduced by polynomial fitting routines; Figure 6.14 shows an example of this.

This error is easily avoided by excluding parts of the image from the fit. This was described in Section 5.1.1. Despite the ease with which this artefact is avoided, it is commonly seen in published AFM images.

6.3.2   Filtering artefacts

Image filtering, by definition, alters the data in the image and therefore always introduces some sort of artefact. When presenting AFM data, it is important to specify what filters, if

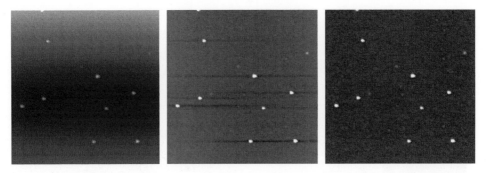

Fig. 6.14. Examples of line-by-line (polynomial fitting) based levelling artefacts. The left image of nanoparticles is unlevelled. The middle image shows an artefact caused by polynomial line-by-line levelling – the particles seem to be sitting in lowered 'trenches' in the background. The correctly levelled image is shown on the right.

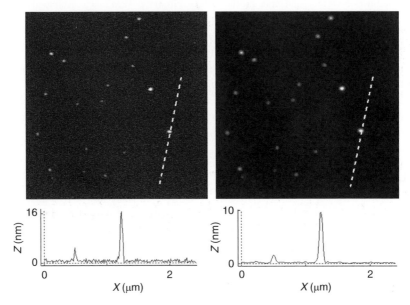

Fig. 6.15. Example of image distortion by filtering. The image of nanoparticles on the left shows considerable noise. Low-pass filtering (smoothing) produced the image on the right. The line profile shows that noise was reduced, but the shapes of the two particles in the line profile were also changed.

any, were applied to the data, because the results from filtered images can be very misleading. For example, low-pass (or smoothing) filters tend to greatly reduce noise in AFM images, but can also introduce artefacts such as changing the shape of features, and increasing the apparent sharpness of steps. An example of filtering artefacts is shown in Figure 6.15.

In addition to matrix filters, as illustrated above, Fourier transform-based filtering can also introduce artefacts into an image. This was described in Section 5.3.4, and shown in Figure 5.12.

6.4   Vibration noise

Environmental vibrations in the room where the AFM is located can cause the probe in the microscope to vibrate and make artefacts in an image. Typically, the artefacts appear as oscillations in the image. Both acoustic and floor vibrations can excite vibrational modes in an AFM and cause artefacts.

6.4.1   Floor vibrations

Often, the floor in a building can vibrate up and down by as much as several microns, typically at frequencies below 5 Hz. The floor vibrations, if not properly filtered, can cause periodic structure in an image. Because it has low amplitude, this type of artefact is most often noticed when imaging very flat samples. Sometimes the vibrations can be started by an external event such as machinery in motion, a train going by, or even people walking outside the AFM laboratory. However, it is often rather difficult to diagnose this type of noise.

6.4.2   Acoustic vibrations

Sound waves (acoustic vibration) can cause artefacts in AFM images. The source of the sound could be from an airplane going over a building or from the tones in a person's voice. The noise of cooling fans from other instruments, or even from the AFM electronics, can also be registered by the AFM. Figure 6.16 is an image that shows the noise derived from a person talking in the same room as the microscope. Diagnosing this type of interference is rather easy; the user must isolate the AFM from the sources of noise or remove them, and look for a change in the signals registered.

The solution to this noise problem, like that from floor vibrations, is isolation from the noise source. Solutions for this were discussed in Section 2.6. Briefly, building vibrations are generally countered by mounting the AFM on a suspended stage that is isolated from the floor. On the other hand, acoustic isolation is accomplished by enclosing the AFM in a cabinet with acoustic shielding on the inside. Alternatively, the noise sources can be removed, and the AFM placed in a location less prone to building vibrations. For this, a room in the basement of the building with little traffic usually serves best.

6.5   Noise from other sources

Floor and acoustic noise are the most common troublesome noise sources in AFM, however, other sources of noise such as electronic noise, which occurs rarely, or noise from a vacuum leak, which is limited only to those instruments that use a vacuum sample mounting system, can sometimes cause problems. The results of poor feedback settings can also appear to give rise to noise in AFM images, when the PID settings are too high.

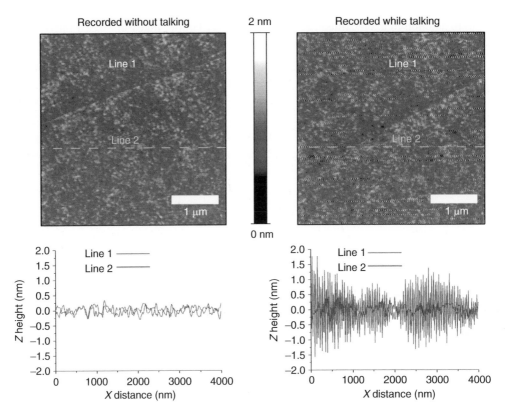

Fig. 6.16. Effect of acoustic noise. This high-z resolution image of a silicon wafer shows the effect of acoustic noise on an image. Right: image and line profiles measured while acoustic noise was present in the room. The acoustic vibrations from a person speaking while the image was acquired are clearly visible in the line scans and the image. Left: image that was measured without the acoustic noise. (A colour version of this illustration can be found in the plate section.)

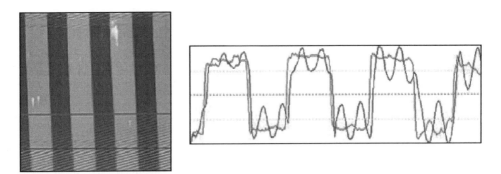

Fig. 6.17. Example of electronic noise in an AFM image. This image of a test pattern has electronic noise at the top and bottom of the scan. The electronic noise in this case was a result of not having a ground wire attached to the stage. The artefact was identified by the oscillation frequency. (A colour version of this illustration can be found in the plate section.)

6.5.1   Electronic noise

Image artefacts can appear in AFM images because of faulty electronics, or accidental electric connections to a part of the AFM. Artefacts from electronics most often appear as regular oscillations or unexplainable repeating patterns in an image, see Figure 6.17. Electronic ground loops and broken components are usually the source of electronic noise.

6.5.2   Vacuum leaks

Atomic force microscopes that are designed for imaging wafers and disks often use a vacuum chuck to hold the wafer/disk while scanning images. A leak in the vacuum between the specimen holder and the specimen can cause image artefacts. The artefact causes a loss of resolution in the image. Cleaning the vacuum chuck and sample and remounting the sample in the stage often eliminates this problem.

6.6   Other artefacts

In this section we gather some other effects that give rise to problems in AFM. Some of these, such as sample drift and surface contamination are the sort of issues encountered in all high-resolution microscopy techniques.

6.6.1   Feedback settings and scan rate

If the feedback (PID) settings used while scanning are not optimized, then it's very likely that the resulting image will show considerable artefacts. This is because the probe is not tracking the surface, and the cantilever is bending to pass over surface features. The correct settings for the PID circuits are also dependent on the scan rate – higher scan rates may require higher PID settings. This artefact can be identified easily by monitoring the error signal. If the error signal is large, then the probe is not correctly tracking the surface. An example of this is shown in Figure 6.18, but see also Chapter 4 for further discussion of feedback parameter optimization. If the PID settings are too high, 'feedback oscillation' can occur, which looks like high-frequency noise in the image.

6.6.2   Surface contamination

As explained in Section 4.1, suitable sample preparation is vital for reproducible, artefact-free AFM imaging. Substantial contamination at the surface of a sample such as a fingerprint or oil film can cause AFM image artefacts. Such artefacts may appear as streaks on the image especially in locations where there are 'sharp' features and edges on the sample's surface. Often the streaking can be reduced or eliminated by cleaning the sample with a high-purity solvent. An example of this effect is shown in Figure 6.19.

6.6.3   Laser interference patterns

Interference patterns can be created by the laser used to detect the bending of the probe cantilever. The interference appears as low-frequency background oscillations in images and typically has a period that is similar to the wavelength of the laser light being used in

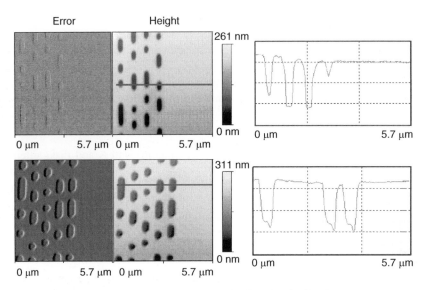

Fig. 6.18. Example of an artefact created by not having the feedback (PID) parameters fully optimized while scanning. In the upper image parameters are optimized, in the lower image parameter are not optimized and the error signal is large. This also leads to less accurate height image, see the line profiles.

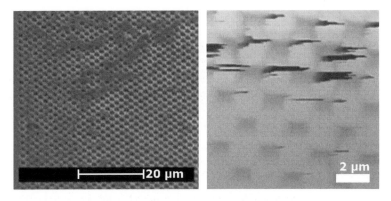

Fig. 6.19. The effect of surface contamination. Left: SEM image of a heavily contaminated calibration grid sample. Right: the contamination causes streaking and prevents the probe from properly following the surface topography in the AFM image.

the AFM scanner (typically 0.5–1.5 microns). This interference originates from laser light spilling over the cantilever, or passing through it, reflecting from the sample surface, and interfering with the light reflected directly from the cantilever. A similar effect can also be seen in force–distance curves, where the interference appears as waviness in the baseline of the force–distance curve, with the same period. This is illustrated along with a typical image showing the artefact in Figure 6.20.

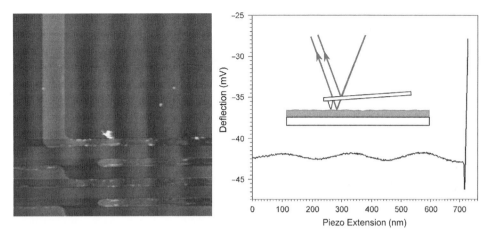

Fig. 6.20. Examples of the effect of laser interference on AFM images and force–distance curves. Left: an image of a reflective sample, showing typical laser interference fringes. Right: the effect on a force curve; the baseline shows similar oscillations. Inset: the artefact originates from interference between the laser beams reflected by the cantilever and the sample.

This effect is reduced in AFM instruments with low coherence lasers, which are fitted in newer instruments. It is also more common with patterned or reflective samples. If the user encounters this problem, it can sometimes be reduced by adjustment of the optical alignment of the AFM. The user should try to ensure the laser is positioned directly in the centre of the cantilever beam, and not too close to the end. See Section 4.2.1 for a laser spot positioning protocol.

6.6.4 Sample drift

A common problem in high-resolution microscopies is sample movement. In general, AFM samples must be well fixed down in order to enable high-resolution imaging. At low resolutions (scans of size larger than 5 μm), some samples do not need to be fixed to the microscope, provided they have a stable substrate. At smaller scan sizes, the sample should be glued to a sample support, which is held (usually magnetically) in the microscope. Even when firmly fixed down some samples can appear to be 'moving' in the microscope. The reason for this is thermal expansion of the sample; this can be exacerbated by sources of heat in the microscope (e.g. the laser or heat from the electronics), leading to samples moving by expansion at hundreds of nanometres per minute, which totally precludes high-resolution imaging. Some samples (e.g. metals) are more prone than others to this effect due to high thermal expansion coefficients.

As shown in Figure 6.21, scanning the sample with the slow scan axis in opposite directions can help to diagnose this problem. Another major problem associated with sample drift is that if the sample drifts in the Z-axis, it can prevent scanning altogether. This can be due to expansion in the Z axis or expansion laterally, which effectively moves the sample in Z, due to sample tilt. Although the feedback system can take account of small

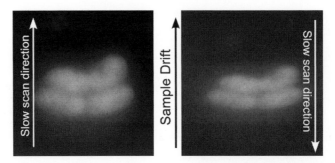

Fig. 6.21. Example of the effect of sample drift on AFM images. The two images of a cluster of *E. coli* bacteria were measured with the slow scan axis in opposite directions. The difference between them indicates that the sample was drifting while scanning. When the sample drifts in the same direction as the slow scan axis, the sample will appear stretched (image on the left); if it drifts in the opposite direction it will be compressed (right image). Scanning in two directions can help to determine the cause of image distortion.

drifts in Z, this effect will eventually cause problems in scanning due to the limited Z scan range of many scanners. If the user determines the sample is drifting, they should attempt to fix the sample down more firmly, and remove possible sources of heat, for example the white light used to illuminate the sample. Sometimes the only solution is to wait for thermal equilibrium.

Chapter 7

Applications of AFM

Although AFM, via STM, had its origins in imaging of metals and semiconductors, it was originally invented in order to be able to extend the possibilities of STM to other samples, in particular biological samples. AFM's ability to image a wide variety of samples, coupled with its simplicity of operation and the relatively low cost of the instruments, means that AFM very quickly gained acceptance in an extremely wide range of fields. In this chapter we arbitrarily divide the illustrative examples into life, physical, and nanosciences, and industrial applications. However, AFM recognizes no such boundaries, and it is hard to think of any field of study involving a solid surface that the technique has not been applied to. The use of these categories can be useful, if only because AFM users have traditionally assigned themselves to one of these areas. However, as illustrated below, the intense interest in nanotechnology and nanosciences has further blurred the distinctions between these areas. This is partly because by definition just about everything AFM studies is defined on the nanoscale, since almost all AFM images have a resolution greater than 100 nm. The range of samples that have been studied by AFM is staggering. Although the areas shown in Figure 7.1 probably cover over 90% of AFM use, AFM has also been used in such diverse areas as art conservation [384, 385], astrobiology [386], geology [387], and food science [388, 389], amongst others.

   The variety of applications of AFM is so large that it's not possible to even mention them all in one book. Instead we have chosen to highlight in this chapter a few applications in each of the categories mentioned above. Mostly, these examples are chosen to illustrate the capabilities of AFM, in particular the different modes and kinds of experiments that can be carried out, rather than to be exhaustive lists of all the applications of AFM to any one field. In each of the four main sections, the introductory paragraph gives an overview of the main applications within the field, and we highlight important advantages of AFM for the field, and more detailed reviews to guide the interested reader to comprehensive summaries of applications within the field.

7.1   AFM applications in physical and materials sciences

The very first applications of AFM were in surface science, and this discipline is still a heavy user of AFM, as well as STM. The ability of AFM to image with atomic resolution is a key advantage for the physical sciences, while the measurement of electrical, magnetic, or mechanical properties also extends the range of possible applications.

   Fundamental applications in physical sciences include imaging of the fine structure of metals and absorbed species on metals and semiconductors; although this is one area where STM is still more widely applied than AFM, some studies show information from AFM can under some circumstances exceed that available from STM. AFM imaging is very suitable for structural studies of inorganic and organic insulators and has been widely

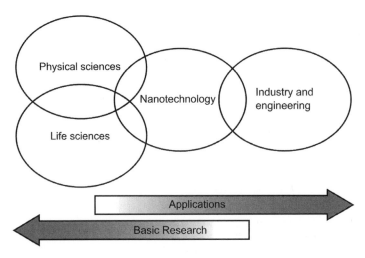

Fig. 7.1. Application areas where AFM is most commonly used.

applied [372, 390–392]. Such fundamental studies of pure surfaces which require atomic or near-atomic resolution are often carried out at low temperatures, in vacuum or both. These types of experiments are outside of the scope of most commercial AFM instruments. Structural studies of molecules have mostly focussed on high-resolution imaging of assemblies of molecules packed into crystals [390], because resolution is generally higher when the molecules are arranged in such highly-ordered structures. However, one of the strong points of AFM is that unlike, for example X-ray diffraction, single molecules can also be studied, and the ability to crystallize the sample is not a prerequisite for AFM imaging. Molecular interactions are commonly studied by AFM, both directly by chemical force microscopy [142] and by studying the topography of the complexes they form [393], and this sort of study is greatly aided by the ability to image in different environments, enabling the study of interactions *in situ*.

The ability of AFM to directly study the sliding of materials over each other (i.e. lateral force microscopy) means the technique is very useful in fundamental work on friction and wear, which is of vital importance in materials science [189, 394, 395]. In fact, AFM is a particularly useful technique in materials science, and other popular applications include the topographic, tribological, roughness, and adhesion/fouling characterization of a wide variety of technologically useful materials [395–397]. Mechanical characterization of materials is also an area where AFM can contribute to their study, especially when studying materials that are heterogeneous on the nanoscale [158, 398–400]. For biomaterials, the modification of the materials by proteins, cells or other biological materials is ideally suited to analysis by AFM due to its ability to image both soft and hard materials [398, 401].

7.1.1 Roughness measurements of high-performance materials

Surface roughness is an extremely important parameter for many material surfaces. Surface roughness can affect adhesion to other materials, optical and electronic properties, surface energy, bioadhesion and other properties [402–405]. Surface roughness is easy to

A B

C D

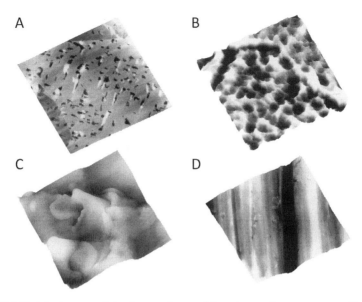

Fig. 7.2. AFM height images of titanium surfaces subjected to various treatments. Samples were treated by A: polishing with colloidal silica, B: polishing followed by acid etching, C: sandblasting then treating with air plasma, and D: grinding with silicon carbide paper. Adapted with permission from [411]. All images have an X-Y scale of 20×20 μm, and a z scale of 6 μm.

measure with AFM, due to the fact that AFM produces high contrast on relatively flat surfaces, and produces three-dimensional, digital data by default. In contrast, scanning electron microscopy often gives the impression that surfaces with roughness values less than 10 nanometres are extremely smooth and featureless, due to the fact that SEM detects electrons scattered from a few nanometres into the surface. On the other hand, optical techniques are limited in resolution, and require opaque surfaces. AFM has no such restrictions in terms of samples, and gives accurate values for surfaces with roughness down to the level of atomic flatness. In addition, neither SEM nor optical microscopy can directly supply data suitable to measure roughness parameters, but determining roughness parameters from AFM data is extremely simple. Therefore, AFM has become the method of choice to measure nanoscale roughness, and is routinely applied to determine roughness of metals and metal oxides [406], semiconductors [407], polymers [408], composite materials [158], ceramics [409], and even biological materials [398, 410].

An example of the use of such measurements is the imaging of titanium, one of the most commonly used materials for many medical implants. Because of its technological importance, and because the surface texture and roughness of the implant is very important for its performance, there have been many studies of titanium and titanium oxide surfaces by AFM [366, 406, 411–414]. For example, in the study by Cacciafesta *et al.*, a series of titanium surfaces with different treatments were compared by AFM [411]. Some example images, of these surfaces are shown in Figure 7.2. The results showed that roughness varied enormously with surface treatment. Clearly, examination of the images shown in Figure 7.2 can give some important information on its own; for example while the polished sample shown in Figure 7.2A displays many pits, it is

Table 7.1. RMS roughness values of the different treatment procedures represented by the images shown in Figure 7.2.

Sample (Figure 7.2)	Treatment	R_q (nm)
A	Mechanical polishing	8 ± 4
B	Mechanical polishing then acid etching	270 ± 27
C	Sandblasting and plasma treatment	900 ± 300
D	Grinding on SiC paper	160 ± 110

considerably flatter overall than the pit-free sample shown as Figure 7.2C. However, the use of the roughness values allows a more quantative approach to surface texture characterization. As an example, the R_q (also known as rms roughness) values of the different treatment procedures represented by the images shown in Figure 7.2 are shown in Table 7.1. It is clear that the processing of these titanium implant surfaces has a huge impact on their roughness, and this is very likely to be an important factor on their performance. For example, roughness values of the surfaces on implants as measured by AFM can be correlated to the degree of bone contact and bone formation of the implant [366, 414]. Rougher surfaces appear to enhance bone contact.

Measuring roughness is a simple and quantitative way to compare surfaces. However, some practical issues must be borne in mind in order to make accurate comparisons of sample roughness. The general rule of thumb is that all scanning, processing and analysis must be identical in order to be able to compare roughness values. The main parameter that affects the value of roughness obtained is the size of the scan measured, so this must always be the same when comparing roughness of different surfaces. On the other hand, the pixel resolution has much less effect on the data obtained, varying only for extreme values [415]. See Figure 7.3 for illustrative plots that demonstrate the relationship between image size (spatially, and in terms of pixel density), and roughness.

Processing of the data, in particular levelling, will affect the value of roughness calculated. For example, with polynomial line-by-line levelling, as is shown in Figure 7.3, higher orders reduce the roughness value obtained [361]. The plot at the right of Figure 7.3 shows a line scan along a femoral hip replacement implant, subjected to different polynomial flattening treatments. The line profiles are clearly quite different, and the values of R_a show that this is dramatically reflected in the roughness parameters. Depending on the reason for measuring the roughness, some levelling should always be applied, so it must simply be applied consistently. Filters should also be used with caution, as they can drastically affect the roughness results obtained.

In order for roughness results to be useful, measurement parameters including the size of the area measured and image treatment should be specified with the results [408, 416]. It has also been shown that the measured roughness depends on the probe used and its condition [408]. This is because less sharp probes will tend to smooth out the data as they cannot image fine features on the sample surface.

7.1.2   Hardness measurement of polymer films

Stiffness or hardness measurement using AFM is one of the more common non-topographic experiments. Although there are a number of other techniques able to make hardness

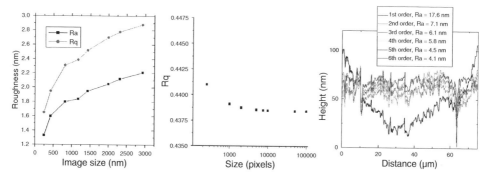

Fig. 7.3. Illustrations of the effects of scanning and data processing parameters on measured surface roughness. Left: the effect of changing the area scanned (image size) on the measured roughness values (R_a and R_q). In general smaller AFM scans show smaller values of roughness. Centre: the effect of changing the number of pixels in the image (pixel resolution) on the roughness (R_q). In general, there's a very weak relation between the pixel resolution and the roughness value. Reproduced with permission from [415]. Right: effect of image processing (levelling algorithm) on roughness of line scans on a femoral head implant replica (R_a). The higher the order of polynomial applied, the lower the roughness [361]. Reproduced with the kind permission of Dr. James R. Smith, University of Portsmouth, UK. (A colour version of this illustration can be found in the plate section.)

measurements of materials, AFM-based measurements have some unique advantages. AFM-based nanoindentation was compared to measurements with a dedicated instrument in Section 3.2.2. The main advantages of AFM are high force sensitivity (hence high sensitivity to differences in sample stiffness, especially for compliant materials), and high lateral resolution, which means that small features or domains can be selectively probed. For these reasons, AFM-based nanoindentation has been widely applied in the physical and materials sciences to probe mechanical properties of micro- and nanoparticles [186, 278], metals [417], silicon [418], and many other materials [419].

Polymers have been a particular focus of AFM nanoindentation studies [168, 181]. One reason for this is that many composite polymeric materials exhibit nanoscale domains. Examples include polymers with fillers or other added particulate materials, and block copolymers. Measurement of the stiffness of such domains can help to understand their contributions to the overall mechanical properties of the bulk materials. In some cases, materials are added to a polymer specifically to change the mechanical properties, such as adding stiffness, or increasing elasticity [420]. Furthermore, the nature of the interface between the reinforcing material and the continuous polymer matrix are extremely important for the mechanical properties of such materials. AFM-based nanoindentation is ideal to probe the mechanical studies of nanoscale phases, as well as their interfaces. Furthermore, measuring the resistance to mechanical probing of a polymer surface can also help to identify individual phases in a composite material [158, 181]. An example of this is illustrated in Figure 7.4. In this case, the material under study was a commercial silicone paint, known to include both large (100–1000 nm) $CaCO_3$ filler particles, as well as small (<10 nm) silica particles. Previous work had failed to detect Ca at the surface, which led to the assumption that features seen at the surface, while of similar dimensions to the $CaCO_3$ filler, had some other origin [421]. AFM-based nanoindentation measurements

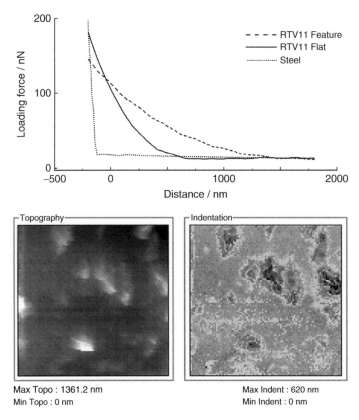

Fig. 7.4. Examples of nanoindentation experiments on a heterogeneous polymer surface. The sample is RTV11, a silicone polymer with calcium carbonate filler particles. The upper graphs show force–distance curves on a flat and a raised region of the RTV11, along with a steel surface for comparison. Below are shown how these experiments may be made in a spatially resolved way. Both topography and indentation images were derived from a map of force–distance curves acquired in a 10 μm square area of the polymer surface. The correlation between the white protruding features in the topography and dark, hard features in the indentation map suggest that these features correspond to the hard filler particles [158].

were made of the polymer surface via a force–curve mapping routine. This technique acquires force curves at user-defined spatial intervals over a user-defined area of the surface. In this case, 100 × 100 curves were acquired over a 10 μm square area, i.e. curves were acquired every 100 nm. Note that even at this low resolution, such an experiment can be quite lengthy – if curves were acquired at 1 Hz, the 10,000 curves would require almost three hours to record. Examples of the force curves obtained are shown in Figure 7.4. It can be seen that there was considerable heterogeneity on the data from different parts of the sample surface. In particular, when measuring curves over the raised features seen on the surface, they were shown to be considerably stiffer than the surrounding polymer matrix, as is seen in the image of indentation distance (Figure 7.4, bottom right). Interestingly, a softer region seems to surround each hard particle, possibly indicating some problems at the matrix–filler interface. These softer interphase domains were approximately 100–200 nm thick, and this highlights a strength of AFM-based nanoindentation for analysis of polymer composites; namely the direct measurement of mechanical property variation at the nanoscale.

The analysis shown in Figure 7.4 was very simple; the distance the AFM indented into the sample was calculated by comparing the cantilever deflection with that obtained on a stiff surface (the steel curve shown at the top of Figure 7.4). Further details of the mechanical properties of polymers can be obtained, however. Thus, it's possible to obtain parameters of the surface such as Young's modulus (E) or sample spring constant, which can be compared to values obtained for bulk materials, or other materials at the nanoscale. However, to obtain such parameters from AFM data, it is necessary to know or to measure the shape of the probe, as well as the spring constant of the cantilever [168]. Methods to measure these parameters were covered in Chapter 2. Furthermore, knowing the shape of the probe, one must model it with an appropriate shape using the measured dimensions, thus the indentation into the surface may be modelled such that E or the sample spring constant can be obtained [419]. AFM-based nanoindentation is a very powerful technique for the *in situ* characterization of the interfaces of heterogeneous polymer systems, and the nature and extent of polymer mixing, as well as other surface and interface effects in technologically important systems can be probed by this technique [168, 181, 422, 423].

7.1.3 Atomic-resolution imaging of crystal structures

One of the most exciting capabilities of AFM is the extremely high resolution that is possible. Achieving atomic resolution, however, is only possible or even useful under certain circumstances. Most solid materials are made up of a diverse collection of poorly organized molecules, meaning that atomic resolution is almost meaningless. We can really only interpret atomic-resolution images in extremely pure samples. For this reason, atomic-resolution AFM is almost entirely limited to application in the physical sciences. However, despite these limitations, some of the results available from such efforts are truly astonishing, such as the work shown in Chapter 3 from Morita and co-authors enabling discrimination of individual atoms using spectroscopic non-contact-mode AFM [8, 424].

Many atomic-resolution images produced by AFM are produced by non-contact-mode AFM (using FM detection), and often carried out in high-vacuum conditions [425]. Thus, these studies are out of the scope of most AFM systems, which need to be specially adapted for FM detection or vacuum work. However, it should be remembered that with great care, ultra-high resolution is also achievable in ambient/liquid conditions [372]. In order to discuss atomic-resolution imaging, it's important to define what is meant by atomic resolution. True atomic-resolution images allow the discrimination of individual atoms. However, some images have been described as 'atomic resolution' that do not allow this, but instead show the *average* arrangement of atoms on a surface. Thus, this *pseudo*-atomic resolution or 'atomic lattice resolution', as we shall refer to it here, does not allow imaging of individual adatoms, or of atomic vacancies. This atomic lattice resolution is rather simple to achieve with a normal AFM under ambient conditions. The main requirements are a sharp, flexible probe, very fast scanning (>20 Hz), and a flat, very clean surface (which can be easily obtained by cleaving mica or HOPG). Two examples of atomic lattice resolution measured with AFM are shown in Figure 7.5. Both of these images were obtained in contact-mode AFM in liquid (water and ethanol, respectively).

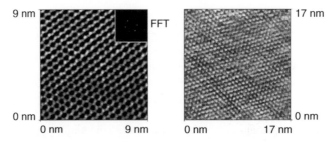

Fig. 7.5. Two examples of atomic lattice resolution. Left: muscovite mica (inset shows the Fourier transform image). Right: self-assembled monolayer on Au (111) surface. These images are not true atomic resolution as no dislocations, point defects or extra surface atoms were seen [372]. Reproduced with permission from [426] and [427].

Images such as those presented in Figure 7.5 allow us to visualize the atomic lattice of crystalline materials, and to make some measurements of it, but their use is limited. The ability to obtain true atomic resolution, and apply it to imperfect or mixed systems is much more useful. This can be achieved under UHV conditions [428] (usually though not always in non-contact FM-AFM), and also in liquid [372, 429, 430]. For instance, this technique allows researchers to observe the atomic structure of a large number of metals, semiconductors, metal oxides [26, 431, 432], and of different materials deposited on top of one another [425, 433]. Crucially, with true atomic resolution, all this is possible with (limited) mixing of components, and allows researchers to image holes, dislocations, stacking faults, sub-monolayers, etc. [372, 434], see Figure 7.6 for examples. This technique can even reveal sub-atomic detail, e.g. information of molecular orbitals can be determined in some cases [435].

True atomic resolution is more routinely achieved in STM than AFM [434]. The main reason for this is that in the imaging mechanism of STM, only the atom on the tip which is the closest to the sample will interact via tunnelling. This means that by manufacturing a 'rough' tip, an atomically sharp probe can be produced. Such an approach does not work for AFM, because even with a very sharp probe, the less specific nature of the interaction means that atoms much further away from the surface than the last one on the tip will also interact with the surface. However, the highly specific nature of the interaction of STM probes with the surface can give rise to some surprising results. For example, because the STM probes the electron density in the surface orbitals, changing the sample bias can make some atoms appear to appear and disappear [436]. A further example is the case of graphite (usually in the form of HOPG). For many years, achieving atomic resolution on graphite has been used as a quality test for both STM and AFM instruments, and it is a very widely studied material by both techniques [428, 437–439]. Some early STM studies appeared to show the expected hexagonal rings of graphite in STM images, but these were later shown to be artefacts caused by double tips [440]. Although in two dimensions, it would appear that all the atoms are equivalent, consideration of the three-dimensional structure of graphite shows that three of the atoms are in one form, named α, and three are in another form, termed β. The α atoms have lower electron density a the surface, and therefore are invisible to STM [441]. Changing the sample bias does not change this. Work by Hembacher *et al.* was able to show that the 'missing' atoms could be seen by AFM [428].

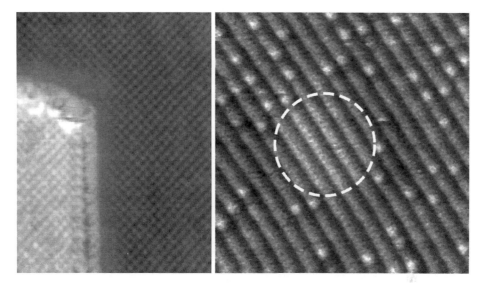

Fig. 7.6. Some examples of true atomic resolution with AFM Left: NaCl islands on a KBr substrate. The interface between the two materials shows artefacts due to convolution with the tip profile. Right: TiO_2 surface. The light rows are O atoms, while the darks rows contain the Ti atoms, and the light dots are OH groups. The dashed circle shows an area of apparently greater height; this is thought to be a patch of charge on the surface. Both images were taken by non-contact AFM in vacuum. Adapted with permission from [433] and [432], both figures copyright (2007) by the American Physical Society.

The experiment was carried out using a custom built instrument that allowed simultaneous measurement by STM and AFM. In order to achieve these impressive results, the experiment was a carried out under UHV, and at low temperature. Simultaneous AFM and STM measurements made using this technique are shown in Figure 7.7. In this case, atomic-resolution AFM was able to give a more accurate representation of the atomic structure than STM was.

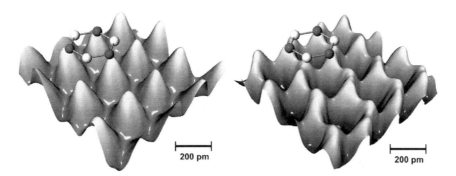

Fig. 7.7. Simultaneously acquired atomic-resolution STM and AFM images of HOPG. Left: STM image with overlaid model of graphite unit cell. Only the β atoms appear. Right: non-contact AFM image obtained of the same region, showing that by this technique all atoms are revealed. Adapted with permission from [428], copyright (2003) National Academy of Sciences, U.S.A.

7.1.4   Friction measurement with AFM

Lateral force microscopy (LFM), or friction force microscopy, is most often applied for one of three purposes: fundamental study of tribology of macroscopic or atomic scale systems [442–444], determination of the friction properties of (uniform) materials [395, 445, 446] or characterization of heterogeneous materials based on their frictional properties [447, 448]. Like some other modes of AFM, LFM allows the determination of parameters that may also be determined by other techniques, which may be simpler to apply or easier to quantify. However, the advantage of LFM is the ability to quantify such parameters on the nanoscale, whether for fundamentals reasons (e.g. the study of tribology at the single atom level), or for practical reasons (e.g. measurement of nanoscopic inclusions in a surface).

Applications include studies of monolayers, including the friction between monolayers on the probe and on a sample surface [444, 448], friction-based discrimination of phases in polymer blends [449, 450] and discrimination of inclusions in heterogeneous polymer surfaces [448], and friction studies of carbide coatings for tool coatings [445, 447]. The ability of LFM to discriminate different chemical groups means it can be used to study phase separation in mixed monolayers [451], and is very commonly used to detect directly deposition of features by dip-pen nanolithography or nanografting, because the contrast is often better than in the height image [260, 452].

Practically, frictional measurements with the AFM are made by measuring friction loops. The term 'friction loop' refers to the combination of the LFM data from the forwards and reverse directions. Two example friction loops are shown in Figure 7.8. The actual friction measured on the material under study is obtained by calculating the difference between forwards and backwards scans. This calculated value is typically derived in terms of volts, but may be converted to force with the methods described in Section 3.2.3.1. The lateral force measured nearly always depends on the normal force applied (e.g. the set-point), and for many materials the relationship will be linear [448], meaning that a plot of normal versus lateral forces allows the calculation of the useful parameter μ, the friction coefficient of the material. An example of such a plot is shown in Figure 7.8.

Measuring quantitative frictional properties as described above is an important application of LFM, but what's really unique about the technique is its ability to distinguish frictional properties of material at the nanoscale. This means that any materials that can be distinguished based on their frictional properties can be differentiated by the technique with the same resolution as contact-mode AFM, *i.e.* up to atomic resolution. For example, 3 nm resolution was seen in a semiconductor film sample consisting of InP and InGaAs regions [455]. As well as the very high resolution, this example highlights another important aspect of LFM for compositional mapping: many materials of very similar composition can be distinguished. Further example are the differentiation of $CdCO_3$ and $CaCo_3$ [456], many different mixed organic monolayer systems [334, 457–459], Si and SiO_2 [460] and many other mixed systems [189]. Figure 7.9 shows an example of how it's possible to use lateral force microscopy to characterize different materials in a heterogeneous surface based on their frictional properties, in this case, filler particles in a polymer film.

7.1.5   Phase imaging to identify surface features

In the first description of phase imaging in AFM [190], It was described how the technique could be used to generate material contrast on a wide range of materials including

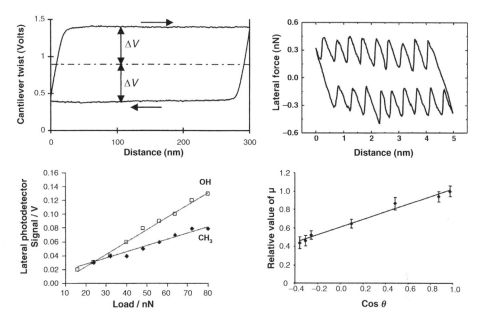

Fig. 7.8. Measuring friction. Top left: friction loop on a homogeneous material. The value of ΔV is characteristic of the materials at a set loading force. Top right: another friction loop acquired on a crystal surface shows atomic-scale stick–slip behaviour. Bottom left: plot showing variation in the friction force as a function of the load for a carboxylic acid-terminated tip contacting hydroxyl and methyl terminated SAMs. From the gradient of these plots, we can obtain μ, the coefficient of friction. Bottom right: the value of μ can be compared to other parameters. Here it is plotted versus cos θ, the water contact angle, which varies as the composition of a mixed SAM is altered. Figures adapted with permission from [336, 453] (copyright 2000 by the American Physical Society), [334] and [454].

composite polymer materials, wood pulp, integrated circuits, quartz/silicon surfaces and more. As described in Section 3.2.3.2, phase imaging is sensitive to viscoelastic properties of the sample and to tip–sample adhesion. This means that many materials can be differentiated by phase imaging, although it is not always possible to directly identify the materials based on their phase signals, and the magnitude of the phase signal is hard to

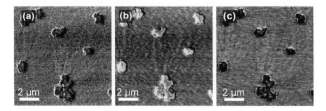

Fig. 7.9 Example of mapping friction changes across a surface to characterize polymer film heterogeneity. The sample is a Mylar D film, including additive particles at the surface. The images show (a) left to right LFM image, (b) right to left image and (c) the subtraction of (b) from (a). The difference in (c) removes topographic effects, and resolves the ambiguity generated by the differences between (a) and (b); the darker colour of the additive particles in (c) shows they have lower friction than the film. Adapted with permission from [461].

predict, as it depends on many factors, including the way the instrument is set up, and the scanning parameters [462].

Because of its ability to distinguish many materials, phase imaging has been applied to an enormous number of samples; just some examples include differentiation of semiconductor films [196], detection of water in cracks in glasses [463, 464], nanoparticle characterization and counting [217, 465], observation of spherulites in polymer crystallization [466], polymer blend and composite composition [467–469] (see Figure 7.10), protein adsorption to biomaterials [411, 470], self-assembled monolayers [462, 471, 472], and many more systems [473]. Due to its dependence on topography, phase imaging is mostly applied to flat surfaces such as films. However, with care, phase imaging can even be applied to image differences in composition on the surfaces of small features such as microspheres [474] or micro-organisms [475].

Phase imaging is particularly useful for imaging dynamic systems, as phase images collected by high-speed AFM are often of better quality than the corresponding topography images [476]. For example, with IC-AFM scanning at high speed often means the z piezo cannot respond fast enough to allow the probe to fully track the sample surface, meaning that height images will be of poor quality. Thus while scanning at high speeds, topography images can show artefacts that are not present in the phase images [477, 478].

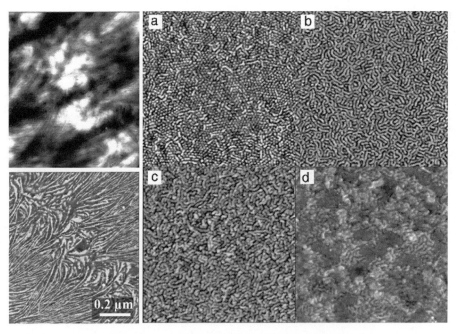

Fig. 7.10. Examples of using phase imaging to characterize heterogeneous polymers. Left: phase imaging is sensitive to crystallinity via viscoelastic properties. In the height image (top left), it's hard to distinguish the lamellae of the polymer due to macroscopic height differences. The phase image (below) shows them much more clearly. Right: example of phase imaging of a nanocomposite. Phase images of sliver nanoparticle/SBS nanocomposite films containing: (a) 0.5, (b) 3, (c) 7, and (d) 10 w.t% Ag nanoparticles. Block copolymers such as this are simply characterized by phase imaging. Reproduced with permission from [466] and [469].

Phase imaging is also sometimes particularly useful for AFM imaging where the sample is dynamic, i.e. for imaging moving systems [472, 479].

7.2   AFM applications in nanotechnology

Without a doubt, AFM is a vital technique for nanotechnology and the nanosciences. The combination of extremely high-resolution, three-dimensional information, and local property measurement means that AFM is considered one of the most important tools in nanotechnology. AFM is used to make important discoveries in all areas of nanoscience. For many systems with dimensions of the order of 10 nm, AFM-based techniques are the only solution to make dimensional, electrical, magnetic and mechanical measurements with the accuracy required [395, 480–482]. In addition, AFM can be used to alter and even to build nanostructures [249, 266]. The examples chosen for this chapter are just a few important examples of the areas where AFM has been shown to be particularly useful.

7.2.1   Nanoparticle measurement

Nanoparticles are probably the most widespread nanostructures, and over the last 10 years or so there has been an enormous increase in the interest in their production due to their relatively simple preparation combined with unique properties. Many of the unique properties of nanoparticles are directly related to their size, for example the photoluminescence of quantum dots which changes significantly if the dots grow a few angstroms in size [483]. For this reason, it is very important to have a tool to characterize with extreme accuracy the size of such particles. Of course, the AFM is also capable of providing more information than just topography, and many other properties of nanoparticles have been probed as well [221, 278, 484–486].

AFM can be used to measure an extremely wide range of nanoparticles including different metal nanoparticles [279, 290, 487–490], metal oxide particles [491], many types of composite metal/organic particles [123, 280, 492–494], synthetic polymer particles [480, 495, 496], biopolymer nanoparticles [283], nanorods [497, 498], quantum dots [499] and others [79, 500]. Some AFM images of nanoparticles with different morphology are shown in Figure 7.11. For pure metallic nanoparticles, TEM is often considered the technique of choice, because metallic particles have high contrast in TEM, don't require coating, and are not affected by vacuum. For such particles, TEM will be just as quick, if not quicker than AFM for measurement of many particles. In Figure 7.11 it is shown how, with care, TEM diameters and AFM particle heights correlate very closely for metallic particles. For hybrid particles with metallic cores and organic coatings, TEM will only show the metallic core whereas AFM will measure both. In these circumstances, therefore, it can be useful to combine the two techniques [280]. However, some particles, such as polymer particles will require extensive sample preparation for TEM imaging (and even then, imaging is indirect), so such particles are better suited to AFM analysis. AFM analysis of coated nanoparticles can even distinguish the thickness of subtly different coatings [493].

Carrying out imaging of nanoparticles for size measurements is exceptionally simple in AFM. Typically, all the analyst needs to do is to deposit a droplet of nanoparticle

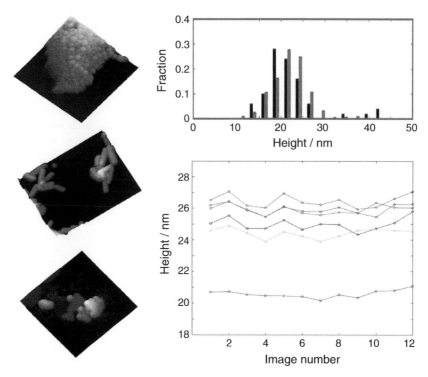

Fig. 7.11. Examples of nanoparticle imaging. Left: example images of nanoparticles with different morphologies; from top: spheres, nanorods [284] and nanotriangles. Right: results from imaging of gold nanospheres. Top: comparison of AFM height measurements (grey) with TEM diameters (black). Bottom: illustration of effect of repeated scanning on apparent height of individual nanoparticles. Each particle was measured 12 times with no apparent change in height. Adapted with permission from [290]. Copyright (2008) Wiley, reprinted with permission of John Wiley & Sons, Inc.

suspension onto freshly cleaved mica, allow it to dry, and image it. The imaging can be carried out in contact, non-contact or intermittent-contact mode, although in some circumstances contact mode may require more care to avoid sweeping effects. The analysis can then be carried out manually or automatically, although in the authors' experience and as reported elsewhere [290], a 'semi-automated' routine works best, i.e. the analyst identifies the particles, and the software routine measures them. This avoids user-dependent analysis, but allows the removal of other features, such as aggregated particles and detritus which can affect the results. More details about sample preparation for nanoparticle analysis are given in Section 4.1, and the image analysis of particles samples is covered in Section 5.3.

7.2.2 *Mechanical measurement of nanotubes*

In addition to their electronic and optical properties, the mechanical properties of pseudo-one-dimensional (1-D) materials such as nanotubes, nanowires, nanorods, nano belts, etc. are the subject of great interest. This is because it has been found that many mechanical

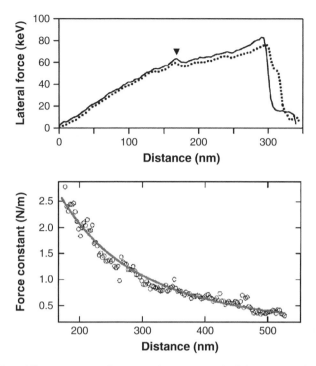

Fig. 7.12. Bending stiffness tests on carbon nanotubes. Top: two individual lateral force–displacement curves, measured at different points along a tube. The small arrow shows the reproducible buckling force. Bottom: CNT force constant calculated from curves as shown above. As expected, the force constant decreases with increasing distance from the fixed tube end, allowing calculation of the material spring constant. Reproduced with permission from Wong *et al.* [506].

properties such as yield strength of one-dimensional materials can exceed those of the bulk materials by orders of magnitude [501]. Indeed, it is due to their extremely high mechanical strength and stiffness that carbon nanotubes (CNTs) are the most thoroughly commercially exploited nanomaterial [502–504]. However, making mechanical measurements of individual nanofibres is extremely difficult, mainly due to the difficulties in locating the objects and fixing them to the testing devices. AFM is ideal for this task, as it can perform both imaging and manipulation of the nanostructures. Furthermore the high positioning accuracy makes it possible to mechanically probe objects having dimensions of less than 10 nm easily [502]. For this reason AFM-based techniques have been widely applied to mechanical measurements of 1-D nanostructures [482]. For example, carbon nanotubes [502, 505, 506], many metal oxide nanowires [507], and metal nanowires [508], and other nanorods [506] have been studied by this technique. Mechanical measurements on 1-D nanostructures using the AFM can be performed in a number of geometries, making it a very powerful technique [509, 510]. Just two methods are highlighted below.

An example of the most commonly used approach for testing 1-D nanostructures by AFM is the direct nanomechanical testing of single CNTs carried out by Wong *et al.* [506] illustrated in Figure 7.12. Both multiwall carbon nanotubes and silicon carbide nanorods were deflected, using lateral force microscopy (LFM) while monitoring the deflection of the AFM cantilever, as the side of the probe tip pushed on the tubes laterally. Before this, the tubes were also imaged by normal AFM, so that both the imaging and mechanical testing capabilities of the AFM were used. One of the major difficulties in this sort of experiment is the problem of how to immobilize the nanotubes while leaving a potion free for mechanical testing [482]. Once testing occurs, the data must be further analysed in order to obtain mechanical parameters of interest, and how the fixed end of the rod is immobilized is important here, too. In the paper by Wong *et al.*, tubes were randomly immobilized above gaps in a substrate, and SiO_2 was used to fix one end of the rod to the substrate. This process is inherently random, so AFM imaging is required to find suitable tubes. The LFM experiments were carried out along the length of the tubes, and it was noted that the carbon nanotubes appeared to buckle at a certain applied force, which was the same wherever along the tube the force was applied. This is shown in Figure 7.12.

Another way to test nanotube structures is by axial compression [505]. This may be carried out in one of two ways; the first method is via attachment to or growing the tube on, the AFM probe, and then pushing it against a solid surface [511]. The second method involves growing a sparse 'forest' of nanotube structures which are grown on or mounted in a substrate, and then uses an AFM probe to compress them by pushing towards the surface [512]. An illustration of the force curves recorder by this technique is shown in Figure 7.13. Application of forces large enough to bend the fibres through almost 180° can be carried out cyclically, as it does not result in bond breakage, due to the remarkable flexibility of both single- and double-walled CNTs [502, 513]. Finally, it should be pointed out that the related two-dimensional material graphene can also be probed by the same and similar mechanical AFM-based techniques, which has produced some spectacular results, notably showing graphene to be the strongest material yet tested [514–516].

7.2.3   Nanodevice construction with the AFM

Ever since Richard Feynman's famous speech 'There's plenty of room at the bottom', a major goal of nanoscience has been the bottom-up assembly of useful devices [517]. Bottom-up assembly means fabricating nanodevices from small components (e.g. atoms or molecules) rather than the traditional ('top-down') approaches of assembling nanostructures by assembly and removal of parts of large components, i.e. lithography.

There are several possible approaches to bottom-up assembly, but these can be broadly grouped into two categories: bulk techniques and nanomanipulation techniques. Bulk techniques are mostly based on careful manipulation of the chemical or biochemical properties of building blocks such that they self-assemble into the desired structures. These sorts of techniques can produce amazing structures, and can produce them in large quantities but the complexity of the devices produced is limited, the manipulator's control over the devices is very limited, and their structures are usually determined indirectly. Nanomanipulation, on the other hand, involves assembling the tiny building blocks directly, one at a time. This approach has the advantage of finer control over the structures formed, and highly complex structures can be formed; on the other hand, it is highly laborious, and is better suited to experimentation rather than mass-production. AFM-based systems are well suited to this sort of task, due to the possibility to move the probe with sub-nanometre accuracy. An additional advantage of AFM for this task is the ability to use the instrument as a sensor as well as the manipulator.

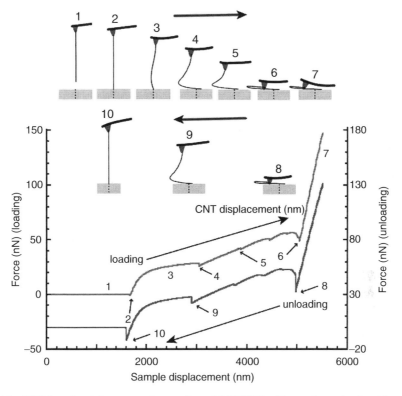

Fig. 7.13. AFM-based axial compression testing of MWCNTs. Top: schematic sketching of the motion of AFM probe and attached nanotube. Bottom: typical result. The grey (upper) and black (lower) traces show loading and unloading response respectively. A remarkable number of features are reproduced in both curves, showing the reversibility of the transitions in CNTs. Reproduced with permission from [505].

There are many ways to perform such manipulations, but usually such manufacture is performed in a semi-manual way, so that the user issues a series of manipulation commands to the AFM, followed by imaging of the results, another batch of commands, etc. One reason for such a laborious manner of working is the fundamental unpredictability of assembly at the nanoscale. Nanoscale objects such as atoms, nanoparticles, etc. do not behave like macro-scale objects, so it is often difficult to predict how they will react to our manipulation [518]. One way to improve the throughput of nanomanipulation is to improve the interface, such that a real-time feedback to the user of the results of their manipulations is possible. There have been various attempts to interface AFMs with alternative sensing systems such as haptic interfaces (allowing the user to feel the sample) [519, 520] or virtual reality (allowing the user to see the sample and/or probe in true 3-D) [520], or combinations of both approaches [521].

For example, an AFM-based nanomanipulator with a haptic interface can be used to move carbon nanotubes onto micron-scale electrodes in order to measure their electrical properties [522]. Similar experiments can be carried out with gold nanoparticles as well

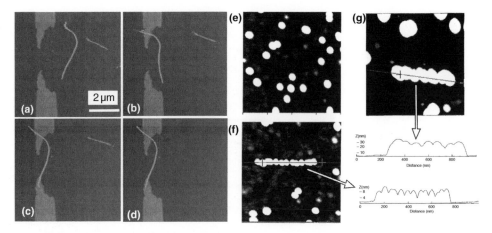

Fig. 7.14. Example of electrical contacts assembled using the AFM as a manipulator. Left: carbon nanotube manoeuvred to connect two gold electrodes. Right: nanoparticles assembled into a nanowire. Reprinted with permission from [522] and [527].

[270, 523, 524]. Many other types of nanostructures have also been manipulated in similar ways, particularly various types of nanoparticles and nanorods [270, 498, 525–531]. In addition, various biological nanostructures (e.g. DNA, chromosomes, etc.) have been manipulated using AFM probes [532, 533]. It should be mentioned that advanced control interfaces make the assembly of complicated devices more convenient, but nanomanipulation is possible using any AFM.

Examples of structures that can be manufactured in this way include wires, transistors [523] and electrical contacts. For example, small gold nanoparticles can be aligned into particle chains by AFM pushing on a surface, and then used as seeds for further gold deposition in order to form a fully connected gold nanowire [527]. Some images showing the process and the resulting wire are shown in Figure 7.14.

7.2.4 Nanoparticle–DNA interactions

One of the most important areas of nanoscience is nanobioscience, which can be defined as using a nanotechnological approach to solve biological problems. Within nanobioscience, probably the most widely applied technology in the medical biosciences is that of nanoparticles, which have been used and proposed for a variety of diagnostic [534], imaging [535] and treatment strategies [536, 537]. One reason why nanoparticles have achieved such broad application is that they approximate the sizes of viruses or even of individual proteins, and thus may interact with the targets as would proteins and viruses. These targets include biomolecules such as proteins and nucleic acids (RNA and DNA), and such targets may be reached *in vitro*, but nanoparticles are also capable of entering cells (as viruses may do) to find their targets [538]. Another reason why nanoparticles are so useful for these sort of applications is that they can be engineered to have multiple properties (such as optical, magnetic, targeting properties) in a facile way [538].

AFM is the ideal technique to observe directly the interaction of nanoparticles (which are usually based on a metal or crystalline semiconductor) with their biomolecule targets;

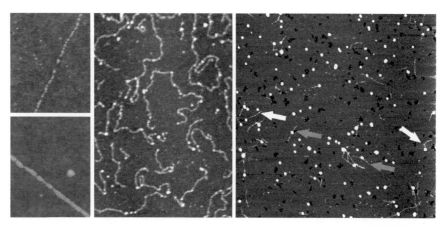

Fig. 7.15. AFM height images of different structures made by nanoparticle binding to DNA. Left: nanowires made by non-specific (electrostatic) binding to 'combed' or straightened DNA; top: sparse nanoparticle coating; bottom: denser coating. Middle: non-specific binding (sugar-DNA phosphate backbone) of glyconanoparticles to DNA [123]. Both DNA and small nanoparticles are visible. Right: probing the action of a biosensor: measuring specificity of binding of DNA-sensing nanoprobes to their targets. Although the nanoparticles are much larger than the DNA, both can be clearly seen. Grey arrows show non-specific binding events, while white arrows show specific binding events. In total, 70% of binding was found to be specific [534]. Reproduced with permission from [547] (left).

in fact no other technique can simultaneously observe both species. There have been a number of examples shown where complexes were observed between various nanoparticles and proteins or DNA using AFM [493, 539–545]. The case of oligonucleotide–DNA complexes is one where AFM can perform a special role. This is because DNA molecules have many potential binding sites for nanoparticles; the AFM's high spatial resolution can be used to distinguish between them. In many examples, the interactions are rather non-specific; the nanoparticles are designed to interact with any DNA sequence. This is particularly useful to build a nanowire using the DNA as a template [542, 546, 547]. However, if the nanoparticles are modified with short sequences of single stranded DNA (ssDNA), they can specifically recognize and hybridize with target sequences in the DNA (or RNA) target [534, 548]. This can form the basis of a biosensor for molecular diagnosis of disease; the mechanism of action of the biosensor can be determined by observation of the complexes of the nanoparticles with target DNA by AFM [485]. Example images showing this are given in Figure 7.15. This is one of a number of methods of specifically targeting certain DNA sequences with nanoparticles that have been illustrated by AFM [534, 549]. Figure 7.16 shows some examples of imaging nanoparticle–DNA interactions.

7.2.5 *Electrical measurements of nanostructures with AFM*

One of the most active areas in nanotechnology is the search for new electronic materials that, based on a bottom-up fabrication approach, could be used to replace today's lithography-based electronics. As well as these applied studies, the study of the fundamental optical and opto-electronic properties of nanomaterials is a very important

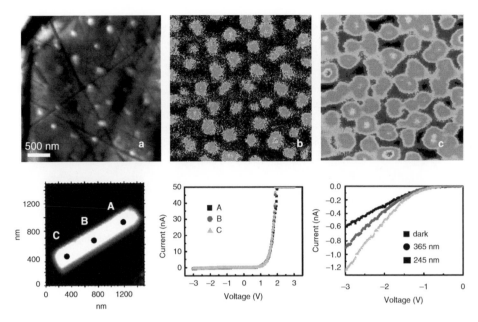

Fig. 7.16. Electrical measurements of nanowires. Top: characterization of a device consisting of vertically aligned germanium nanowires in an aluminium oxide substrate. The surface was polished and the aluminium selectively etched away to expose the tops of the nanowires which can be seen in the topography image (a). Conducting AFM with a platinum-coated probe was used to measure the current passing thought the nanowires, while potential differences of 20 V (b) or 40 V (c) was applied between the probe and the other ends of the wires. All the wires seen in the topographic image are conductive, and some that are not obvious in the topography can be seen in the current images. Bottom: electrical measurements on ZnO nanorods as a function of distance along the rods, and of light flux. Left: AFM height image of a ZnO rod showing the locations where *I-V* curves were measured. Middle: *I-V* curves at different places – the response is uniform, indicating no defects. Right: *I-V* curves measured on an individual nanorod during dark conditions, and during illumination at two different wavelengths, showing the frequency-dependence of the device. Adapted with permission from [565] and [563].

area of nanoscience. All these measurements require that electronics contacts can be made to the nanosized components, typically with placement resolution on the order of 1 nm. This is a major experimental challenge.

Conducting AFM techniques are ideally suited to overcome this difficulty of electrical measurements of carbon nanotubes, one of the most important materials for nanoelectronics [550–554]. Directly measuring the properties of individual CNTs is made much simpler by the ability of AFM to position an electrode (i.e. a conducting AFM probe) at any point along the tube desired. AFM can also be used to deliberately introduce defects into pristine CNTs in order to measure the effect on their electrical properties [553, 555–558]. Other nanostructures that have been probed electrically include quantum dots [559, 560], many types of nanowire [561–565], nanowire-based transistors [566], and even electrically active biological nanostructures such as single metal-containing proteins [567–570], and other types of nanostructures [571, 572].

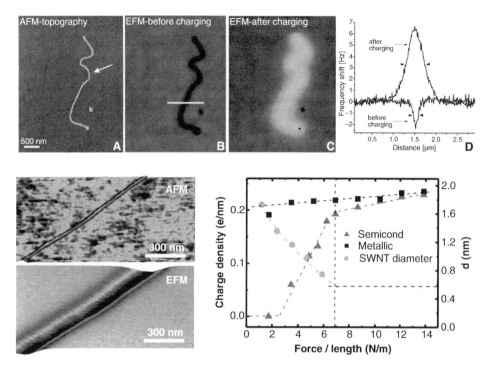

Fig. 7.17. Electrical measurements of carbon nanotubes. Top: charge injection into nanotubes. **A**: topography of isolated MWNT, and its EFM image before charging (**B**). After charge injection by the AFM probe (at the arrowed location), the EFM contrast reverses(**C**). **D**: a line profile through the EFM images at the point indicated by the white line. Bottom: measurement of the effect of mechanical compression on charging behaviour of SWNTs. Left: AFM and EFM (measured by lifting) images of a SWNT after charge injection by the AFM tip. Right: plot showing charge density λ (red triangles) as a function of compression of the semiconducting CNT by the probe (indicated by the diameter, d, which decreases as the CNT is compressed, green circles). The black squares show the behaviour of a metallic SWNT, illustrating how compression with the AFM changes the behaviour of the semiconducting SWNT to match the metallic behaviour. Adapted with permission from [550] and [558]. Lower figures copyright (2008) by the American Physical Society. (A colour version of this illustration can be found in the plate section.)

AFM has unique advantages for the electrical characterization of carbon nanotubes, specifically the ability to combine high-resolution imaging with electrical characterization both by imaging and spectroscopy of electrical properties, at the single-particle level. An example of the use of AFM techniques in the characterization of carbon nanotube electrical properties in shown in Figure 7.17. In the upper portion of the figure, the effect of charging on the EFM signal from CNTs is given. The leftmost image shows a standard IC-AFM height image of an isolated MWNT. EFM was carried out in a lifting mode using conducting probes, and the initial result is shown in the second image, which shows a negative frequency shift of the cantilever over the CNT, indicating attractive force between the probe and CNT. Charge was then injected into the nanotube using the AFM probe, by approaching the tube with a bias applied between the sample and the probe.

The EFM image after this process is shown in the third image. At this time, both the tip and CNT are negatively charged, and thus a repulsive interaction occurs, leading to a positive frequency shift. Line scans over the CNT in the EFM images are shown on the right. It is interesting to observe that although the charge was injected at one point on the CNT (indicated by the arrow in the left-most image), the nanotubes is homogeneously bright in the EFM image, except for a small defect at the end, showing the spreading of the charge through the nanotube. This effect can be used to detect inhomogeneities and defects in CNTs [550].

Another advantage of AFM for this sort of study is that with AFM it is possible to mechanically challenge the nanostructures in addition to electrically changing and probing them. The relation between the electrical properties of CNTs and their mechanical deformation is a topic of intense interest due to potential applications [573]. There are several different ways in which AFM can be used to alter CNTs mechanically, and then measure the effect on their electrical properties [550, 556, 558, 574]. In the example shown in the lower part of Figure 7.17, two types of SWNTs are considered, semiconducting and metallic nanotubes. The semiconducting nanotubes are able to maintain only a small charge in EFM experiments similar to those shown at the top of Figure 7.17. However, upon compressing the semiconducting nanotubes mechanically by the AFM probe (shown in the figure by the green circles, which is the measured diameter of the tubes), the behaviour of the semiconducting CNTs change, and approaches that of the metallic CNTs (shown by the red triangles approaching the black squares). In this example, AFM was used to change the sample, as well as to characterize the effects both in terms of topographical and electrical changes. This allowed the direct observation of mechanically induced semiconducting–metallic behaviour crossover which had been previously only indirectly observed [558, 575].

7.3 Biological applications of AFM

The biological or life sciences constitute without a doubt, one of the most important application areas for AFM. This is evident from the fact that nearly all AFM manufacturers build specialist models of AFM for biological sciences, and there even exists at least one company that *only* makes AFMs designed specifically for applications in the life sciences. This is despite the fact that *any* AFM instrument can be used for biological applications. In fact, AFM itself came about partly in order to extend the possibilities of STM to biological samples. Many innovations in AFM technique and instrumentation which are now used in other application areas also came about due to the interests of biologists in the use of AFM, such as IC-AFM, and later the extension of IC-AFM mode to use in liquid [576], and low-noise force spectroscopy.

As a microscopy technique, AFM has several key advantages for biological application. Probably the most important of these is the ability to work under physiological-like conditions. Almost all biological processes occur in liquid, and often depend strongly on the presence of certain salts, and the temperature of the solution. Many biological samples also change their structures dramatically when dried. Therefore the ability of AFM to image and measure samples in buffer solution, at 37 °C, at any ionic strength or pH is of vital importance to many biological experiments. Furthermore, AFM is particularly simple to integrate with optical techniques which are very important for many experiments in the life sciences, such as epifluorescence microscopy or confocal microscopy. Combining these advantages with the possibilities that AFM offers to carry out other experiments, such as mechanical probing, molecular interaction measurement by force spectroscopy, etc. means that AFM is an extremely important tool in many biological areas.

For these reasons, the number of AFM applications in enormous. The sections below include only a few selected areas where AFM has proven particularly useful, but more applications in the life sciences have been covered elsewhere [577–579].

7.3.1 Biomolecule imaging

Biomolecules form the basis of life, and understanding the structure, function and inter-actions of biomolecules has been the key to the incredible progress in the life sciences, and medicine in particular, over the last 50 years. As a technique with sub-molecular resolution and the ability to image soft samples in water, AFM is very appropriate for the study of the huge range of natural biomolecules. The four major classes of biomolecules are carbo-hydrates, proteins, nucleic acids and lipids.

Of these, probably the least-well studied by AFM is the class of carbohydrates, although even here, a number of different systems have been studied. These include self-assembled monolayers of glycoconjugates [9, 580], glycosylated particles [280, 581], polysacchar-ides [582, 583] and force spectroscopy of carbohydrate interactions [144, 584].

On the other hand, proteins have been extensively studied by AFM, not just by imaging but also by other measurements, such as electrical and mechanical measurements [320, 567, 585–587]. Due to their importance in disease and biological processes, proteins are one of the most widely-studied classes of molecules, and literature searching reveals thousands of studies of proteins using AFM. A few representative examples will be given here.

AFM is a particularly suitable technique for protein studies, due to the coupling of high resolution with the ability to study samples under physiological conditions, which is necessary because protein structure can be highly sensitive to the nature of the proteins' environment. However, despite the incredible resolution achievable on flat atomically well-defined surfaces, AFM of single proteins in physiological-mimicking conditions often gives rather low resolution, only revealing sub-molecular features for very large proteins or multi-domain protein complexes. This is due to the soft, yet tightly packed and globular nature of most protein structures, meaning an AFM image of the outside topography shows few structural details. Indeed, it is extremely challenging to obtain angstrom-level resolution of native proteins under such conditions by *any* technique because of the fact that the molecules are under constant movement.

However one area where AFM has been used to provide great details is in protein complexes. Such assemblies are often studied by TEM or crystallographic techniques, which suffer in that they study the complexes under of non-realistic conditions, giving AFM an obvious advantage in the fidelity of the data to biological systems [301]. One such protein complex that has been quite widely studied by AFM is the GroEL/GroES chaperonin complex. This complex is of interest because of its role in assisting the protein folding process [588, 589], and has been quite widely studied by both contact and oscillating AFM modes in buffer solutions [99, 590–592]. Sub-molecular resolutions, namely distinguishing the seven sub-units of the oligomer can be achieved [99, 590]. The GroES ring acts like a 'lid' for the GroEL, sitting atop it under some conditions, and the GroEL without the lid can be distinguished from the entire GroEL/GroES complex [590], allowing the kinetics of the association/dissociation of the two partners to be studied by high-speed AFM [304, 593], see Figure 7.18.

A major class of protein assemblies that have been extensively studied by AFM is that of fibrillar assemblies, which have great importance for human health, due to their implication in diseases such as Alzheimer's disease and their importance in blood clotting [301]. For example, the formation of protein nets and fibrils formed during blood clotting has been observed on mica and HOPG surfaces [594]. Collagen, the most abundant protein in mammals, also forms fibrils. These have characteristic band structures, which are simple to observe by AFM, and this has become a standard specimen for biological AFM [35, 595–599].

Individual protein monomers can also be studied by AFM, although it is sometimes difficult to image isolated molecules without adhering them to a surface [306]. One preparation method commonly used for membrane proteins is to image them in a phospholipid bilayer, which stabilizes them towards imaging, as well as mimicking biological conditions. Proteins in lipid bilayers are covered in Section 7.3.3. Covalently binding proteins to a surface can stabilize them toward single-molecule imaging, although this might affect the protein structure. In some cases, careful control of pH and ionic strength is enough to enable imaging of isolated proteins absorbed onto mica [301, 306].

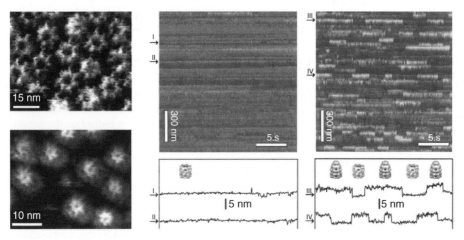

Fig. 7.18. AFM imaging of single protein complexes. Left: high-resolution images (contact-mode AFM) of GroEL (top) and GroEL/GroES assemblies (bottom). The association of the GroES onto the GroEL adds about 5 nm to the height of the complex, and covers the internal cavity. Right: high-speed imaging (IC-AFM) of GroEL/GroES association and dissociation. The images use a novel 1-D imaging technique to increase speed. The slow scan axis was disabled, and the fast scan axis (vertical) repeatedly scans a line containing various molecules. In the centre image (in the absence of ATP), this leads to continuous stripes along the time axis (horizontal), as little association or dissociation was occurring. When ATP is added (right image), the assembly heights switch rapidly (see line scans below), indicating rapid association and dissociation. In both examples the images were measured in buffer solutions. Adapted with permission from [590] and [304].

Increasing the concentration of protein or their aggregates on a surface can lead to increased stability and imaging resolution, as at high surface concentrations the proteins can stabilize each other. The ultimate extension of this is to form two-dimensional crystals at a surface, which can lead to greatly improved imaging quality, and has been shown to allow sub-molecular resolution in some systems [102, 600]. However, despite the impressive resolution this technique is applicable only to some proteins, and removes one of the great advantages of AFM, i.e. single-molecule imaging and relevance to biological conditions [100].

Nucleic acids are highly suited to AFM imaging, and can be imaged in air and in liquid, and by contact, non-contact and intermittent-contact modes [300, 319, 601]. Samples have included single stranded DNA (ssDNA), double stranded DNA (dsDNA), and even the unusual triple stranded DNA [602]. RNA in both the common single stranded form (with tertiary structure) [603, 604], double stranded RNA [605, 606], and ssRNA in an extended configuration have been imaged [479]. Complexes of proteins (usually enzymes that have the nucleic acid as a substrate) with both RNA (anti-RNA antibodies-RNA complexes [605]) and DNA (e.g. DNAse-DNA complexes) have been imaged in air [607].

High-quality images of DNA can be obtained by deposition of a solution onto freshly-cleaved mica, followed by imaging by NC- or IC-AFM. Reproducible and high-resolution images require some way of binding DNA to the mica, because both mica and DNA are negatively charged under common conditions. This is usually done by either treating the mica with a divalent cation solution before deposition, or including such a cation (e.g. Ni^{2+} or Mg^{2+}) in the deposition buffer. The divalent cations are thought to act as a salt bridge [319], and this treatment allows imaging in air (after washing away most of the salts, followed by drying), or in liquid (the imaging liquid must then contain the divalent cation). Alternative methods include treating mica with an amino-terminated silane [300, 302], although caution must be taken not to increase greatly the roughness of the mica by this method, as imaging the DNA well requires a very clean surface. See Chapter 4 for more sample-preparation details.

One great advantage of the electrostatic absorption via divalent cations, is that if imaging in liquid, careful control of the ionic conditions in the imaging buffer can ensure the DNA stays on the surface, while allowing it freedom to move in two dimensions, and even to carry out physiological functions [2, 4, 478, 608]. This has led to some very elegant experiments in which both DNA and proteins are 'bound' to a surface well-enough to be imaged by AFM, while being free enough to carry out their interactions in real-time. Of course, such molecules bound to a mica surface are not under true physiological conditions, but no other technique allows molecular biologists real-time single-molecule imaging of these sorts of reactions at all [609]. Some stills from a time-lapse 'movie' that can be generated by this technique are shown in Figure 7.19. More applications of AFM to studies of DNA are discussed in the review [610].

The IC-AFM images shown in Figure 7.19 were acquired at a rate of 1 image per second. This imaging rate is remarkably fast for AFM, especially for IC-AFM images of such delicate structures. However a long-term goal for fast-AFM imaging researchers is to improve the speed of IC-AFM data acquisition even further, allowing AFM to probe protein–nucleic acid reactions with high time and spatial resolution [476].

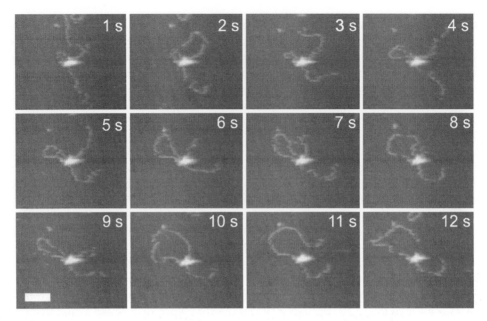

Fig. 7.19. An example of real-time measurements of DNA-protein interactions. These images were obtained with high-speed IC-AFM, and the numbers in each frame indicate the time after imaging began. The images show a complex of EcoP15l, a DNA restriction enzyme, with DNA. The images from 1 to 10 seconds show the DNA loop passing through the enzyme. Reproduced with permission from [2], copyright (2007) National Academy of Sciences, USA.

7.3.2   *Bacterial cell measurements*

AFM is a highly suitable tool to examine bacteria, and has been widely applied to their study. Bacteria are commonly studied by optical microscopy, which can give an overall idea about gross cell morphology (via a two-dimensional projection), and is also useful for cell-counting studies. In comparison, AFM is slower, and thus is less useful for quantitative cell-counting, but allows measurement of a variety of other cellular properties, particularly by nanoindentation and force spectroscopy experiments [611]. In addition, the greatly increased resolution of AFM allows for the imaging of finer details of cell morphology and sub-cellular features such as pili and fimriae [612]. The three-dimensional information from AFM can also be useful in differentiating morphologies which would look the same in optical microscopy [6]. Various other micro-organisms have been studied by AFM such as spores [178, 613–615], fungi [616, 617], including yeasts [171, 618], viruses [287, 619], and others [620] but here we concentrate on bacteria for the sake of brevity.

Some species of bacteria that have been well-studied include *E. coli* [169, 621] and various species of *Staphylococcus* [169, 317, 622], *Bacillus* [178, 615], *Streptococcus* [623, 624] and *Salmonella* [625, 626], see Figure 7.20. Bacteria generally need to be immobilized on a surface for imaging, and a number of different procedures have been used. For studies in air, drying onto a surface, or even flaming can work well, although one

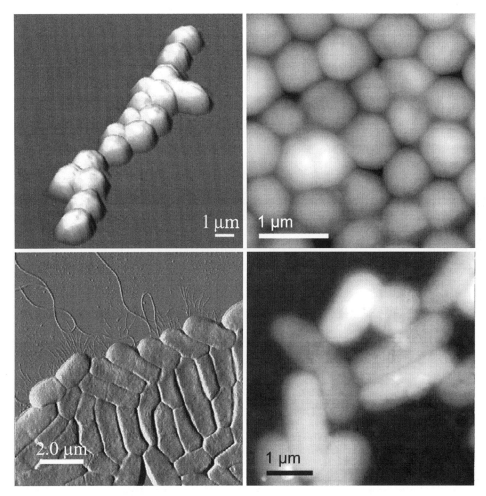

Fig. 7.20. Studies of bacterial morphology. Top left: *Streptococcus*, showing typical linear clusters. Top right: large clusters of *Staphylococcus aureus*. Bottom left: *Salmonella* biofilm showing pili-like fimbrial structures. Bottom right: *E. coli*. All these images were measured in air. Reproduced with permission from [624] (top left) and [626] (bottom left).

needs to be careful of drying artefacts by these techniques [621]. However, it is useful to be able to study bacteria in liquid, and usually this requires a more elaborate preparation protocol because it's necessary that the cells be fixed to the substrate in some way. The most commonly used techniques include the use of poly-l-lysine (PLL) or polyethyleneimine (PEI) coating of glass for chemical capture and using gelatin-coated glass for a soft physical capture [6, 314, 315]. For spherical bacteria, i.e. cocci, physical trapping of substrates with appropriately-sized holes works well, and it can even be possible to observe the cells dividing while immobilized in this way [313, 317]. Bacteria that naturally form biofilms are simple to study as biofilms are perfect samples for AFM, although some washing may be required. See Section 4.1 for more sample preparation details. Although higher resolution is usually obtained in air, bacteria imaged in liquid are closer to the native state, and dried bacteria usually have a small fraction of their hydrated height [6, 315, 621].

One of the most important areas in studies of bacteria is the study of the method of action of antibiotics and other antibacterial agents, due to the ongoing increase in antibiotic resistance in bacteria [627]. Several studies have imaged bacteria treated with antimicrobial agents, including the morphological changes to *E. coli* caused by the antibiotic cefodizime [628] and also *E. coli* and *P. aeruginosa* response to antibacterial peptides [629, 630], *S. aureus* response to antibiotics [631, 632] and others [624, 633, 634]. The response to the natural antimicrobial polymer chitosan, of *E. coli, S. aureus, B. cereus* and *B. cereus* spores has been measured by both AFM imaging and nanoindentation measurements [169, 178]. The changes that can be seen include morphological alterations such as appearance of holes, shrinking, cell shape changes and cell lysis, and also mechanical changes. An example showing the response of *S. aureus* to antibiotic treatment by both topographic changes and changes in cell elasticity is shown in Figure 7.21.

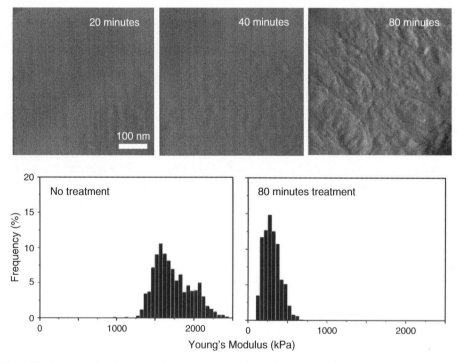

Fig. 7.21. An example of the use of AFM to measure bacterial response to antibiotic treatment. Top: topographical images showing cell wall roughening in *S. aureus* after increasing times of treatment by the antibiotic lysostaphin. The images are deflection images measured in contact mode. Below: the effect on cell wall stiffness. The drug appears to severely degrade the cell wall within 80 minutes. This data was collected in buffer solution, on a cell trapped in a membrane pore. Reproduced with permission from [632].

It's also useful to make measurements of bacteria by non-imaging modes of AFM, because the high positioning resolution of AFM allows such measurements to directly address individual bacterial cells, which is difficult by other techniques [611]. For example, nanomechanical measurements (e.g. nanoindentation) of bacteria have been shown to be sensitive to treatment with antimicrobial agents [169, 629, 632], bacterial species and strain [155, 635, 636], physiological state of the organisms and the environment in which the measurements are made [155]. With the AFM it's relatively simple to perform nanoindentation experiments on individual micro-organisms, and even to differentiate one part of a cell from another by stiffness measurements [171]. For this sort of experiment, it's important to remember that the response of the probe will be different when the cell surface is perpendicular to the probe motion, than when it's at an angle, however [637]. Thus, all measurements should normally be carried out only on the upper portion of the cell which is relatively flat [382].

Other non-imaging experiments which may be carried out on bacteria using AFM include force spectroscopy in order to measure the distribution of specific adhesion factors on cell surfaces [156], cell hydrophobicity/hydrophilicity [475, 638], or the distribution of other molecules across the cell surface [611, 637, 639].

Bacterial colonization of surfaces is an important process, and reducing the process requires knowledge of individual bacteria–surface interactions. Bacteria–surface adhesion studies can be carried out using a number of experimental methodologies, the most commonly applied ones being direct force spectroscopy with bacteria immobilized on the AFM probe and lateral force microscopy measurements of the force required for removal of cells [637, 640–644]. AFM allows the combination of studies of cell–surface adhesion, with measurements of the surface itself, which can help to understand how factors such as roughness, hydrophobicity, etc. can affect colonization by bacteria [645].

7.3.3 Lipid membrane imaging

Plasma membranes are ubiquitous in animal cells, forming a barrier between the intracellular components and the extracellular environment. The membrane's purpose is to selectively allow molecules in and out of the cell, while blocking unwanted material. In addition the membranes form a scaffold for a large number of cell surface molecules, mainly proteins, which regulate activities such as cell adhesion, recognition, signalling, etc. By AFM it is possible to study the cell membrane in its native environment, i.e. as part of a cell (see next section), but for increased stability and higher resolution, it's useful to use a model system. The major component of the plasma membrane is a bilayer of phospholipids, with their hydrophilic heads pointing out into solution, and the hydrophobic tails on the inside of the bilayer, so that they are shielded from the aqueous environment. Due to their importance in biology such lipid bilayers have been widely studied by a number of techniques. While they are simple to study by other techniques in solution (they form spherical vesicles), for AFM they can be deposited easily on a flat surface. This creates a flat, stable model for the plasma membrane, which is an ideal sample for AFM studies.

Formation of lipid bilayers for AFM studies has been carried out using a number of different methodologies. The two most common of these, however, are Langmuir–Blodgett film deposition, and fusion of vesicles from solution directly onto the substrate surface.

Lipid membranes form extremely flat surfaces, and membrane proteins or lipid domains often are visible in AFM as sub-nanometre height features, so it is generally desired that the substrate be atomically flat. In order to form a bilayer the substrate should be hydrophilic, so freshly cleaved mica is typically the best substrate to use, although silicon [646] or template-stripped gold [647] may be used, particularly when a specific surface chemistry is required. Langmuir–Blodgett (LB) film deposition is carried out by first forming a layer of the lipids on a water surface, and passing the substrate through the air–water interface [648]. This has some advantages in terms of control of the pressure of the bilayer, and the ability to form half-bilayers or mixed bilayers. On the other hand, vesicle fusion is an extremely simple process. A solution of vesicles is pipetted onto a mica surface and typically left for 20–40 minutes for fusion to take place, before rinsing with buffer solution. The vesicles collapse on the surface, leaving a well-organized bilayer, with typically some remnant vesicles which are washed away [646, 649]. This preparation process is quick and simple to carry out, although it lacks some of the control of the LB film technique. An advantage for imaging under liquid is that the sample need never dry out during the preparation process. Imaging of bilayers can be carried out in contact or oscillating modes, (either non-contact AFM or IC-AFM) [649], [125] and in air or, more commonly, in liquid [650].

In either case, when bilayers are prepared, the sample is often referred to as a supported lipid bilayer (SLB). This is to make explicit the fact that these bilayers are not in vesicle form, but on a surface; they will therefore have some interaction with the substrate surface. In fact it's known that the interaction with the surface changes somewhat the properties of the bilayer compared to a vesicular bilayer. An example showing height differences in the bottom bilayer of a multi-bilayer stack due to this surface interaction is shown in Figure 7.22. Studies of bilayer structure can include measurement of phase separation in mixed lipid systems, which is an important process due to the involvement of lipid rafts in many biological processes [129, 651]. Phase separation can be studied by a number of AFM techniques. Due to the incredibly high height resolution of AFM, discrimination of phases with a few Å height difference can be carried out directly [129]. In addition, friction contrast (measured by LFM) [652–654], or phase imaging [471, 655] can help to differentiate phases of very similar heights. Examples of this are shown in Figure 7.22. If a nanoindentation-type experiment is carried out on lipid bilayer with a flexible probe, then at a certain threshold force, a 'breakthrough' can be observed in the force–distance curve [129, 656]. This can be used as a measure of the coherence of the lipid film, and its thickness may be measured from the curve. Practically, this measurement can be used to prove the existence of the films, as lipid bilayers on mica can be uniform to the point of being featureless [308].

AFM is also ideal to study the interaction of peptides or proteins with membranes [657]. Some peptides are known to disrupt and damage lipid membranes, whilst other are drug candidates that need to be able to cross membranes. The changes in lipid membranes are typically disruptions with dimensions on the order of a few nanometres, so the action of these materials is usually studied indirectly. With AFM their action can be observed directly, typically manifesting as appearance of small 'holes' or other morphological changes in the SLBs [658–661].

Probably the most important application of studies of SLBs is the study of protein incorporation in membranes. By AFM this is rather simple, and can be performed by

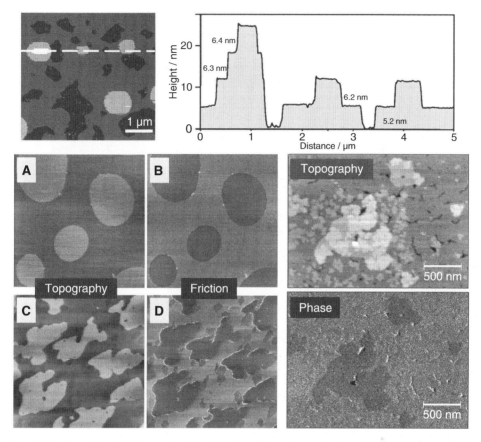

Fig. 7.22. Examples of bilayer structural studies. Top: demonstration of the influence of the substrate surface on bilayer height. The four layers are identical in composition, but the one next to the substrate appears considerably thinner. Below: differentiating phases in phospholipid mixtures by LFM and phase imaging. Left: topography (A,C) and LFM (B,D) of a mixture of DSPE and DOPE. The monolayers were imaged in air (A,B) and under liquid (C,D). Right: phase imaging of *E. coli* total extract in liquid. Note that there are several different lipids in this mixture, and some are discriminated by phase imaging, while others are not. Reproduced with permission from [129] (top), and [652] (left).

simply fusing proteolipsomes instead of liposomes onto a mica surface [308]. The protrusion of the membrane proteins above the flat lipid surface can then be imaged directly, and can give important information about protein insertion, see for instance the example on the left of Figure 7.23. Lipid membranes have also been used as a scaffold to study the fine structure of membrane protein oligomers in their native state. In AFM, membrane proteins have been widely studied in 2-D crystals, which are a highly stable configuration, allowing high-resolution imaging [320]. However such systems do not very closely represent the native conditions for membrane proteins. On the other hand, incorporation of protein complexes into membranes allows their imaging in conditions much closer to the native ones, and can demonstrate different structures to those found in the

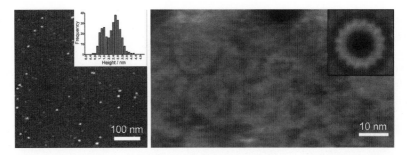

Fig. 7.23. Examples of proteins and protein oligomers inserted in lipid membranes. Left: low-resolution imaging of single outer membrane proteins (OmpF) inserted into DMPC supported lipid bilayer[308]. The inset histogram shows two peaks in the histogram of protein heights suggesting two insertion mechanisms. Right: high-resolution images of light-harvesting complex incorporated into a DOPC/DPPC SLB. The inset shows an averaged image of the complexes, showing the discrimination of the 16 sub-units. Reproduced with permission from [662] (right).

crystallized complexes, which are presumably closer to the native structures. AFM is the only technique to allow high-resolution imaging of protein complexes in near-native conditions [662]. An example showing high-resolution imaging of protein complexes inserted into an SLB is given in Figure 7.24.

7.3.4 Mammalian cell imaging

Due to their importance in biological and particularly biomedical sciences, animal cells have been widely studied by AFM. As a high-resolution microscopy technique able to image samples under physiological conditions (in buffer or growth medium, at controlled temperature, with controlled ionic strength), AFM has some unique advantages for cell biology. In addition, the ability to make mechanical/chemical measurements using the AFM probe enables further possibilities. AFM is also particularly easy to combine with

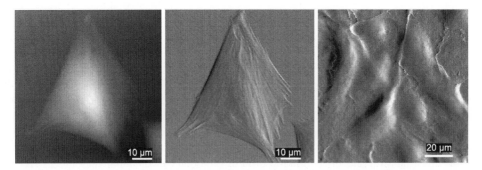

Fig. 7.24. Examples of AFM imaging of live animal cells. Left: contact-mode images; height and deflection images of osteoblast cells on a polystyrene culture dish (height image z-scale: 3.9 μm). Right: IC-AFM image (amplitude image) living fibroblast cell layer on PLL-coated glass slide. With cells, error signal images such as the deflection and amplitude images shown here are usually presented as they show more details. Reproduced with permission from [678] and [103].

optical microscopy, so overlaying florescence images with AFM data is possible, allowing the combination of the unambiguous identification of features by florescence labelling and the high-resolution imaging of AFM. However, it is also true that imaging living cells presents the AFM operator with some unique challenges. Notably, very high resolution, such as may be obtained regularly with other samples can be extremely challenging to achieve when imaging living animal cells [103]. This is likely to be due to a combination of the living cell's high sensitivity and flexibility. Living cell surfaces are able to move spontaneously in solution [663], thus establishing with certainty their location with a mechanical technique is difficult.

However, while requiring great care, high-resolution imaging is possible, and even the comparatively low-resolution images that may be attained more routinely have resolution many times greater than optical microscopy [101]. In imaging mammal cells, the first decision to be made is whether to image the cells live, in solution, or dried and fixed. This will typically depend on the particular application, and the information required. While imaging dried and fixed cells can give useful information [664], live cells in liquid will be less prone to fixation artefacts, and closer to native conditions, and can also enable imaging of dynamic processes [665]. The second decision is whether to image the cells in contact or oscillating modes.

Despite the mechanical softness of living cells in solution, contact-mode imaging can give surprisingly good results [101], and is probably used more commonly for live cell imaging than oscillating modes [103]. Typically contact-mode imaging is carried out with very small applied forces and very soft cantilevers (spring constant, $K < 0.1$ N/m) [101, 666]. For oscillating modes in liquid, slightly stiffer levers are usually used to overcome probe–cell adhesion. Unlike most other applications, for live cell imaging, it may be preferable to use unsharpened silicon nitride probes, which have relatively large tip radii ($ca.$ 20 nm). These may be less likely to penetrate the cell membrane, leading to higher achievable resolution than with sharper probes [667].

Because of the flexibility of the cell membrane, different imaging conditions can lead to different images. Typically with contact-mode imaging (especially at high applied force), sub-membrane features (such as the cytoskeleton, actin fibres, etc. [668]) are visible, while for oscillating modes, the membrane itself is shown and the cytoskeleton is not seen [103, 669]. In addition, changing the applied force can affect the visibility of sub-membrane features. While increasing the applied force to make more features visible, can cause greater apparent resolution, it can also result in sample damage [101, 666, 670]. A few examples of contact and IC-AFM imaging of live cells, illustrating the differences commonly seen, are shown in Figure 7.24. One important aspect of cell imaging for AFM can be seen in the figure: animal cells are very large samples for AFM. While high-resolution imaging of the cell membranes can be very useful, it is usually convenient to also obtain overview images showing whole cells such as seen in Figure 7.25. This requires a large scanner, with a wide X-Y range (>50 μm, preferably 100 μm), and some cell types necessitate a long Z axis travel ($ca.$ 10 μm) as well – this is particularly the case for measuring cell–cell or cell–substrate adhesion [671, 672].

Some of the more common applications of cellular imaging include morphological studies, which includes the morphological changes in cells upon interaction with drugs or other biological molecules [668, 673, 674], observation of dynamic cellular processes [675, 676], and observation of morphological changes in diseased cells [665, 677]. Due to

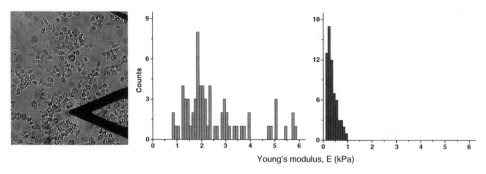

Fig. 7.25. Using AFM to diagnose cancer – nanomechanical measurements of live cells. Left: optical view of an AFM cantilever probing mixed healthy and tumourous cells. Middle: Young's modulus of healthy cells. Right: Young's modulus of tumour cells. The cells were differentiated by immunostaining. The AFM-based nanoindentation results show that the tumour cells are much less stiff, enabling cell motility in the body. Adapted with permission from [680].

the high-resolution imaging, and also the ability to image not only the cell, but it's surroundings, studies of the interactions between cells and their substrate are a strong point of AFM-based cell studies [175, 678]. For example via AFM it's possible to observe cell responses to nanostructured surfaces, and even to observe the cell–surface interactions in real time with living cells [664].

Mechanical properties of animal cells are very important for their functions, and thus it is useful that AFM can measure cellular mechanical properties at the single cell, or sub-cellular level [668, 679–681]. It is possible to determine differences in stiffness between different parts of the same cell, typically showing the presence of sub-membrane cell components. One example of the utility of this is in probing the mechanical differences between diseased and healthy cells. Recently it was realized that an important factor leading to tumour invasion and metastasis is a decrease in cell stiffness, leading to increased ability to spread throughout the body [682]. AFM has been used to prove that tumour cells collected from patients are considerably less stiff than healthy cells from the same patients. Further, it appeared that various types of cancer cell, all from different patients exhibited very similar stiffness values [681]. The AFM-based measurement of stiffness was proposed as a diagnostic method for forms of cancer that are difficult to diagnose by traditional methods [680, 681]. In addition to nanoindentation experiments, force spectroscopy experiments have been widely applied to living mammalian cells, which are covered in the next section.

7.3.5 Biological force spectroscopy

Intermolecular interactions are the basis of life, and an extremely important part of biological research, so an enormous range of techniques have been applied to their study. For the interactions of biological molecules, AFM has some unique advantages. It is very sensitive, allowing the interactions between single pairs of molecules to be studied due to force resolution in the 10 pN range [683]. Moreover, it can be very selective. The control over the x, y and z position of the probe with immobilized molecules means that unlike most techniques, by AFM-based force spectroscopy, it's possible to control exactly which molecules interact, and where they are doing it. This is fundamentally different from bulk techniques, where the molecules of interest are placed in a relatively large volume, and some signal change observed. In addition, because the molecules are brought together and pulled apart under the control of the experimentalist, factors such as force of interaction and the rate of separation can be finely controlled. Combining all this with the possibility to carry out such reactions in physiological conditions means that AFM-based biological force measurements represent one of the most powerful techniques available to make molecular interaction measurements in biology [684].

There are two major ways in which force spectroscopy can be carried out: in one or in three dimensions. One-dimensional (1-D) force spectroscopy refers to experiments where the factor of interest is the intermolecular force to be measured, rather than the spatial distribution of the measured forces. In order to carry out this sort of experiment, the AFM probe and a flat surface will be modified to bind the two molecules of interest. The flatness of the substrate will reduce artefacts related to increased adhesion at the edges of features [159]. For self-interaction, the same molecule could be bound to both the surfaces [144, 685]. There are a number of issues related to the binding strategy used, which were discussed in Section 3.2.1, but essentially, the molecules must have a resistant yet flexible linker, and not have their recognition sites blocked [686, 687]. When suitably modified probes and samples have been generated, experiments are carried out as described in Section 3.2.1, and any pairs of interacting molecules may be studied. A common way to prove the nature of the interaction being probed is to add a 'blocking' molecule to the solution [163]. For example, having established that with molecule A on the tip and molecule B on the surface, a measurable interaction force is recorded, molecule B is added in excess to the medium. These excess 'B' molecules bind to 'A' on the tip, and the force spectroscopy experiments are repeated. If the interaction being probed is really of type 'A-B', the measured force will be changed (often it will disappear), with the blocking molecules in solution. The main parameter which is studied is the force or range of forces, at which detachment occurs (hence *force spectroscopy*). However, for molecules bound by a flexible linker, or for macromolecules, the distance of unfolding is also important (see next section). Typically the rate of pulling (i.e. the speed at which the probe is moved) can be varied, and this can also allow measurements of the kinetics of the dissociation process [688–690].

Some examples of interactions that have been studied include biotin–avidin binding [163, 691, 692] (which has become a 'standard measurement' in AFM force spectroscopy, due to its very high strength and specificity [690, 693]), other antibody–antigen interactions [689, 694, 695], carbohydrate–carbohydrate binding [144, 584], and fibrinogen binding [685, 688, 696], see the examples in Figure 7.26. AFM can also be used to measure cell–cell adhesion, or virus–cell adhesion by attachment of a cell or virus to the probe [697, 698]. This is only a small selection of the interactions that could be studied; for further details see the reviews [142, 684, 699].

Three-dimensional force spectroscopy or force-mapping is the second major methodology of force spectroscopy by AFM. In this sort of experiment, the aim is to use the specific force of interaction between two molecules to determine the location on the sample surface of one of the molecules. For instance, this can be used to determine the location of receptor molecules on a cell surface. Probably the most commonly used way to do this is by measuring force–distance curves in a grid pattern over the sample surface. This is sometimes referred to as force volume imaging [700]. In this mode, the

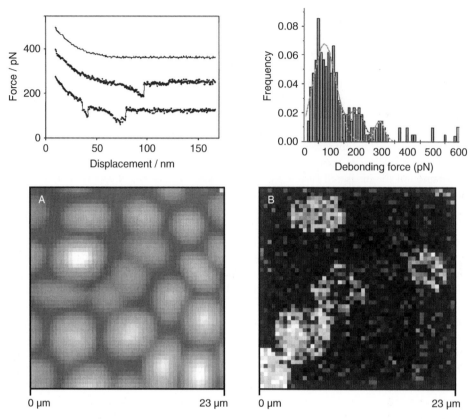

Fig. 7.26. Examples of biological force spectroscopy. The two examples show different methods of studying interactions with human blood cells – at the top, one-dimensional force spectroscopy, and below, three-dimensional force mapping. Top: force spectroscopy on human platelets. The top force–distance curve was made with an unmodified tip, and the bottom two with a tip modified with peptide sequences from fibrinogen, showing the results of single (middle curve) and multiple (lower curve) adhesion events. On the right is a 'force spectrum', showing the presence of peaks at multiples of *ca.* 93 pN. Below: force mapping on red blood cells with a lectin-modified probe. Image **A** is the total adhesion force and image **B** is the topography of a mixed layer of group A and O cells. The topography shows no difference between the cells, while the adhesion image clearly distinguishes 'A' from 'O'. Adapted from [688] and [710].

AFM instrument approaches the probe to contact the surface, and then pulls the probe away, before moving a small amount while out of contact, approaching again, etc. This is done in a grid pattern defined by the user. This method of measuring interactions has great advantages in that the individual force curves are as well-defined and controllable as via normal (1-D) force spectroscopy, and are carried out normal to the sample surface. However, it is also a rather slow technique, and normally the maps are produced with reduced resolution (e.g. 64×64 points [142]), in order to make the experiments reasonably short. In addition, the data processing can be complicated and time-consuming.

Force mapping in this way has two major advantages over more commonly used approaches to determine molecular distributions in biology, which usually involve labelling the receptor. Firstly, no labelling, which could affect the results, and requires prior knowledge of receptor chemistry, is required, and secondly, the resolution is higher than optical techniques. It can be used for a very broad range of applications, principally the mapping of the locations of various molecules on the outer membranes of live cultured mammalian cells [672, 701, 702], yeasts [703], bacteria [156, 704, 705], etc.

In addition to force-mapping via force–distance curves, it is possible to carry out similar experiments using dynamic AFM modes, with the main aim of increasing the speed and resolution of force mapping. This is sometimes known as dynamic recognition imaging or affinity imaging [706]. This can be done in a type of intermittent-contact AFM, using a probe modified as for force–distance curve acquisition [707]. The interaction between the modified probe and the targets on the sample surface should alter the probe's response in IC-AFM, and the signal corresponding to this change must be extracted from the data measured while the probe scans over the surface. One way to do this is by oscillating the probe (with an amplitude lower than the length of the linker between the probe molecules and tip), and electronically extracting two signals from the measured probe oscillation – that from the upper part of the oscillation (the oscillation maxima) and that from the lower part (the oscillation minima). The lower oscillation signal is for feedback. This is the part of the oscillation most affected by the mechanical damping of the probe by the sample surface. The upper part of the oscillation signal is used for the recognition image. This part is sensitive to the interaction between the probe-absorbed molecules and those on the sample surface [708]. The chief advantage of this technique over force–distance curve mapping is that molecular recognition can be recorded simultaneously with topographical imaging – at the same speed as normal AFM imaging. The trade-off for this speed is a reduction in the amount of information available at each point (for example the pull-off force, chain extension length, etc. which can be measured directly when using force-mapping). Such dynamic techniques may be applied to molecules absorbed to flat surfaces [709], or even to receptors on cell surfaces [685].

In addition to making measurements of intermolecular binding, force spectroscopy can measure the strength of intramolecular bonds, for example the force required to separate the two strands of dsDNA [711], to unfold the secondary structure of ssRNA [712], or to unfold proteins, which is covered in the next section.

7.3.5.1 Protein unfolding

Measuring protein unfolding with AFM is an advanced application of force spectroscopy. However, because protein unfolding is a huge area in biophysics and biochemistry, the adaptation of AFM to measuring protein unfolding created a whole new field of experiments [713, 714, 715, 716], and it is becoming an increasingly common application that has led to improvements in experimental technique in force spectroscopy and in instrumental capabilities and force resolution [586]. One reason that protein unfolding by AFM is so interesting is that in the classical techniques, protein unfolding is induced by either chemical or thermal denaturing. While these are important pathways, the ability to induce unfolding via completely different mechanisms allows researchers to probe the process in a very different way, revealing aspects of the unfolding process previously inaccessible [717, 718]. Furthermore, certain proteins require tensile strength for their physiological

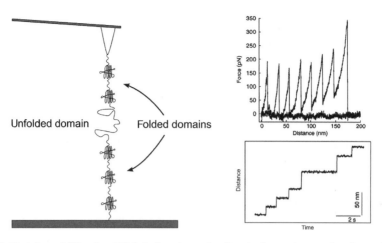

Fig. 7.27. Protein unfolding by AFM. Left: schematic of typical experiment, showing polyprotein being stretched between AFM tip and a surface to which it is covalently bound. Top right: typical force curve measured on a polyprotein in constant velocity mode. Bottom right: typical result from the same sample in constant-force mode. Adapted with permission from [586].

function, and thus mechanical resistance is an important part of their design [719]. While there are other techniques which may be used to study mechanical unfolding of proteins (such as optical tweezers), the accessibility and simplicity of protein unfolding by AFM has meant it is the most popular technique with which to study the phenomenon today [717]. The folded structures of proteins are typically held together by forces such as hydrogen bonding, hydrophobic, ionic, and van de Waals interactions, which are all weak interactions, but are collectively strong enough to hold the structure together. Therefore, the fine details of unfolding pathways require high force resolution. The way in which traditional force spectroscopy works is not ideal for protein unfolding studies, because normal force spectroscopy is carried out at a constant velocity. This means that in protein unfolding, large changes in applied force will occur during the process. A method that was developed to overcome this limitation is force clamp spectroscopy, in which an additional force feedback loop is added to the instrument, to maintain a constant force during unfolding [720, 721]. A further derivative of this technique is force ramp spectroscopy, where the feedback loop is used to maintain a constant increase in force during unfolding. At the time of writing, nearly all force-clamp spectroscopy has been carried out with modified instruments [722], but commercial instruments with such capabilities have also begun to appear. Typically such experiments are carried out with synthesized polyproteins, large molecules with multiple copies of a single protein domain [716, 718]. Pulling this type of molecules should give rise to a characteristic 'fingerprint' force curve (i.e. the sawtooth- or staircase-shaped profiles shown in Figure 7.27), which helps to reduce the ambiguities in force spectroscopy data discussed above. A new generation of commercial instruments have been produced recently, which are designed to optimize the ultimate force resolution by reducing noise in the z axis, largely spurred by the requirements of protein unfolding experiments. It has been estimated that using standard commercial cantilevers, thermal noise limits the resolution to approximately 6 pN [723]. These novel experimental methodologies and instrumental improvements have led to new insights into how protein structures control their physiological functions [717].

7.4   Industrial AFM applications

Presently the majority of AFM applications are in basic and applied research at universities, government and large company laboratories. However, there is a growing trend in the use of AFMs for commercial applications. The growth depends on the following requirements:

(a) Images and data from the AFM must be reproducible.
(b) The instruments must be operable by technicians and users that are not experts in AFM.
(c) There must be an awareness of AFM techniques at all levels in a company from bench level to senior managers.

For the purposes of this book AFM applications are divided into two categories: applications associated with commodity products, and applications for high-technology products. One of the limitations in presenting commercial applications is that companies often do not permit publication of their images and applications, because the information is considered confidential. It is not until many years after the application is identified by a company, that it is made public. Due to this, there are relatively few references for industrial work.

The following sections provide some insight into how atomic force microscopes are used in an industrial setting. It is by no means a complete list, as many applications are being carried out in complete secrecy, and cannot be presented publicly.

7.4.1   Commodity product applications

Commodity products are made in extremely high volumes with processes that are for the most part very well understood. Typically the products have very low profit margins per unit but can be very profitable because they are made in such large volumes. Because of the nature of this type of business, expenses for product development and research account for a very small percentage of a company's operating budget, and AFM has been used less often in this sort of industry than in the high-technology sector.

It is interesting to note that many commodity products have historically had critical components that were nanometre sized. Without knowing it, highly controlled processes were developed to manufacture of products having these nanometre sized components. Car tyres are an example. A car tyre is constructed mainly from rubber and carbon black. The carbon black is a nanoparticle. The quality of a tyre is related to the size of the carbon black particles and the dispersion of the carbon black in the rubber.

The motivation for companies to use atomic force microscopes when they are selling commodity product is typically to improve existing products. The goal is to make the products more cheaply, better, faster, stronger, etc. There are many consumer products now having claims of improved performance because of nanosized components. Examples include tennis rackets, golf balls, hair products, clothes and washing machines.

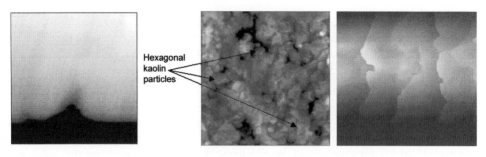

Fig. 7.28. Examples of commercial products with applications in AFM. Left: 4 μm × 4 μm image of a razor blade. Middle: image of clay-containing coating on paper, showing hexagonal kaolin particles. Right: 35 μm × 35 μm image of human hair.

Another motivation for companies that make commodity products to use the AFM is to solve production problems. As an example, a company that makes photocopy machines was having problems with the machines breaking because the particles used in the reproduction process were not working correctly. By using an AFM the company was able to find that the particles purchased from another company did not meet their specifications. Before it was solved the problem was costing the company approximately 1 million dollars per month.

In addition, AFM analysis can be used to support advertising claims, or to resolve industrial disputes. As an example, a toothbrush manufacturer was claiming that their product caused less damage to teeth than their competitors. The competitor challenged this claim with a law suit. An AFM was then used to compare the wear on teeth from the two companies' products. In fact, it was demonstrated that the claims were true.

Figure 7.28 shows a few examples of commercial samples scanned with an AFM. The razor blade image is an example of using AFM to characterize defects caused by faulty processes. Razor blades are sharpened to atomic dimensions in extremely high volumes. The AFM was used to profile the sharpened surface and view imperfections at the very edge of the blade. Mounting the razor blade in the AFM so that the scans could be made without causing further damage was a substantial challenge.

Secondly, a paper coating is shown. The quality of paper coatings greatly affects their performance, and many papers are complex composite materials, containing different mineral particles, elastic binders, and fibres in the surface structure. For centuries paper products have been made from nanoparticles. A common component which controls the quality of the paper is clay. Often, the higher the clay content, the higher quality the paper. An AFM can readily be used to visualize the clay nanoparticles in paper. The image in Figure 7.28 shows an AFM image of Astr-Plus/Carbinal 95 coating. This 'engineered' coating is comprised of kaolin in a narrow particle size distribution and an ultra-fine ground calcium carbonate with a latex binder. AFM is a highly suitable tool for imaging and quantitative analysis of paper coatings [724, 725].

The final example of a commercial sample for AFM is hair treatments. Hair care products can cause substantial changes to the overall geometry of human hair. The changes can affect the hair's optical and frictional properties which are very important to consumers. AFM is the only method that is able to measure the three-dimensional topography of human hair, and in addition can make quantitative measurements of hair surface roughness and frictional properties [726, 727]. AFM can be applied to image the effects of many different cosmetic products on hair, for instance, the effect on topography and friction of hair bleaching [728], or the effects of shampooing and conditioning with different products can be studied [729].

7.4.2 High-technology applications

High-technology companies produce high profit-margin products that usually use advanced manufacturing technologies. These types of companies commonly adopt new techniques such as AFM to facilitate their research efforts, and less often, may use AFM to support production when necessary. The digital data storage industry and the semiconductor industries rely on AFM for solving some of their most difficult problems. Other products that have used AFM for product development include flat panel displays, optics, micro-electromechanical systems (MEMS), and biosensors.

High-technology industries require novel materials, devices and processing capabilities. In a commodity-based company product development is 5% of revenue whereas high-technology companies spend 10–15% of their revenue on product development. Thus high-technology companies can afford to invest in atomic force microscopes.

Such industries often require high dimensional tolerance. Further, high-technology companies need to have an understanding of fundamental material behaviour in order to improve performance, manage quality control, and to improve product yields. An atomic force microscope is one of several tools available to high-tech companies which are capable of making such measurements. It should be noted though, that the AFM is a relatively new tool and typically other microscopes, such as electron microscopes, are used before an AFM for applications in technology industries.

Another characteristic of technology companies is that they often have to improve and change their processes in a very short time period when compared to traditional commodity-based companies. Thus, a company may need to use an AFM to support a manufacturing process for only a few years, after which the company may not need the microscope.

The limiting issues for broad acceptance of atomic force microscopes in the high-technology industries are probably probe quality and image capture rates. Variations in AFM probe geometries result in unwanted variations in quantitative measurements. Additionally, if a probe is damaged while scanning a surface, it is difficult for an operator to know, and erroneous data may be collected. When compared to optical and electron beam techniques, the AFM is sometimes considered to be slow. Thus, the AFM is used only for product development and off-line production applications.

Specialized AFM products are often required for high-technology industries. For example, in the semiconductor industry, the AFM must be particulate free so that it does not contaminate product wafers. Also, sample sizes can be very large in the high-technology industries. In order to design an AFM that can handle large sheets of glass for the LCD industry, the final instrument must be very large and heavy. It is not uncommon for these AFM instruments to weigh several thousand pounds. For many applications the AFM must be able to automatically measure several images on the same sample.

7.4.2.1 Semiconductor industry

The use of AFM in the semiconductor industry started in the mid-1990s. Since that time several hundred atomic force microscopes have been employed by semiconductor companies. The majority of the applications were never made public. The applications presented here are some of those that were made public. Most of these applications were associated with developing new processes, processes that most likely have a lifespan of 5–10 years.

Front-end (wafer fabrication) applications include measurement of surface texture and dimensional metrology. The polishing of silicon wafers in the planarization step of semiconductor manufacture can be studied by AFM. The tolerance in these manufacturing processes is very low, and the signal-to-noise ratio of the measurements must be very high, which is why for such applications only AFM is suitable. The noise floor of the AFM must be below 0.5 Å for adequate characterization. The planarization method of chemical–mechanical polishing (CMP) was developed with the aid of AFMs for quality control. The second major application in semiconductors is in metrology. The measurement of feature and trench dimensions is very important, and due to its high resolution and three-dimensional information, AFM is a powerful tool for this [730]. Specialized probes such as those in the shape of a pole, which reduce ambiguity at feature edges, or even with flared ends that can measure line edge roughness have been developed for these applications [731]. Such measurements do not require measuring an entire image but only a line profile over the trench. The time required for the measurement is greatly reduced when only a line profile is required. Atomic force microscopes give excellent contrast on extremely flat surfaces, and so are very useful for thin film characterization. Because the AFM creates a 3-D map of the surface, software algorithms for measuring grain sizes are very reliable. In the case of insulating films, there is a great advantage of AFM over SEM because the sample does not need to be coated with a conductive film.

Back-end applications (product assembly, packaging and testing) in the semiconductor industry include solving problems associated with packaging, thermal management, adhesion of contacts and bonding. The thermal measurement capabilities of the AFM have been used to help solve heat transfer problems in several generations of microprocessors. Defects associated with problems in production processes can be investigated in detail with AFM. The defects can sometimes be seen with an optical microscope, but the AFM image can give greater insight into the source of the defect. Direct electrical measurements with the AFM can also help in such cases. Examples of these semiconductor applications are given in Figure 7.29.

7.4.2.2 Data storage

Advances in the data storage industry occur at a staggering pace. Data densities of all types of media increase dramatically on an annual basis. Atomic force microscopes were first used by the data storage industry to study the magnetic domains on hard disk drives in the early 1990s. At the same time, atomic force microscopes were used to improve and study optical disk drives. This has continued until today, and AFMs played a critical role in the development of CDs, DVDs and their successors. Custom stages are available to hold disk masters, stampers and replicas. With automated software it is possible to make statistical measurements of bit dimension distributions. Common measurements are the bit width,

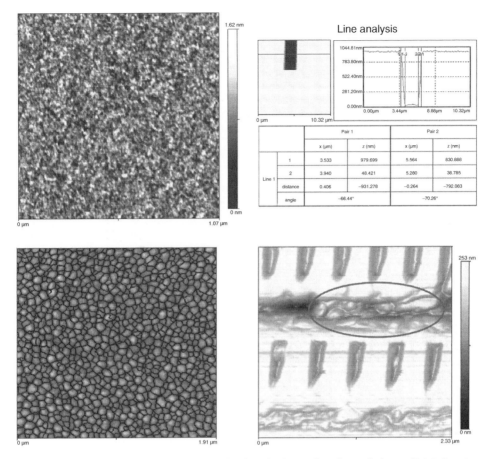

Line analysis

		Pair 1		Pair 2	
		x (µm)	z (nm)	x (µm)	z (nm)
Line 1	1	3.533	979.699	5.564	830.888
	2	3.940	48.421	5.280	36.785
	distance	0.406	-931.278	-0.264	-792.083
	angle		-66.44"		-70.26"

Fig. 7.29. Examples of high-technology applications in the semiconductor industry. Top left: using AFM to measure the quality of polishing by CMP. The CMP processes can achieve extraordinary flatness, as may be seen from the very small z-scale of this height image. The rms roughness (R_q) of this image is less than 1 nm. Top right: metrological measurement over a trench on a patterned wafer. Bottom left: thin film characterization by grain analysis on a polysilicon film. Bottom right: example of a defect imaged in a cross-sectioned device (circled).

length, and the angles of the bits sides. Although the AFM is helpful in developing these products, the AFM is typically not used in the mass production of optical media. A very common application for the AFM in the data storage industry is the study of pole tip recession on hard disk drive read/write heads [732]. The manufacture of hard disk heads is another example of a process that has extremely low error tolerances, due to the small head–platter distance in hard drives. AFM is used to accurately measure the recession (trench) that is used to protect the poles in operation [733]. Of course, MFM is also capable of measuring the magnetic domains with high resolution, which becomes increasingly important as data density increases [734]. Another application in the hard disk industry is measurement of laser bumps. These bumps on the surface of hard drives are used to

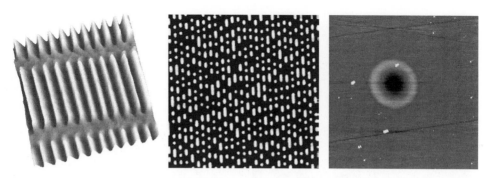

Fig. 7.30. Examples of applications in the data storage industry. Left: Three-dimensional representation of magnetic domains, in a 24 μm MFM image of a hard disk drive, allowing measurement of domain structure and spacing. Centre: 20 μm height image of a DVD. Automated software allows measurement of many properties from images such as this. Right: 16 μm height image of a laser bump in a hard disk landing zone. The scratch-like features are burnish marks, intentionally applied to allow air to escape while the head flies over the disk platter.

facilitate the landing of the drive head when not in use. They are created in the landing zone of the disk surface with a short laser pulse. The bumps are not very deep and cover a small area of the disks surface. The AFM helped with the development of the process to create these pits; the height of the pit's lip as well as the depth of the pit can be directly measured with an AFM. Some examples of these applications from the data storage industry are shown in Figure 7.30.

7.4.2.3 MEMS devices

One of the first commercial applications for MEMS devices was the cantilevers/probes used in the atomic force microscope. Applications for the AFM in MEMs devices include measuring the surface roughness of reflective surfaces, measuring the forces required to move MEMS fabricated devices, and metrology measurements on MEMS devices. Figure 7.31 illustrates a device imaged by both SEM and AFM, showing how the true

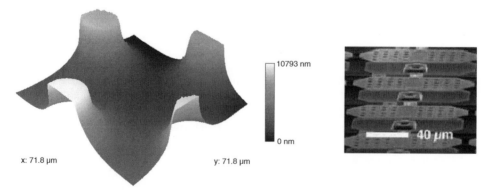

Fig. 7.31. Example of imaging of a MEMS device. The AFM image, on the left, gives quantitative height information, while the SEM image (right) does not.

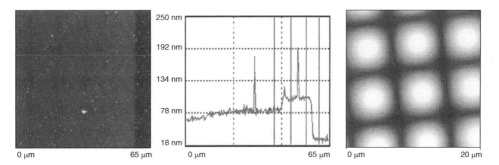

Fig. 7.32. Optical and video products: on the left is an image of an LCD screen, where the step height is characterized with a line profile. On the right is a micro-lens array, consisting of a surface on which a series of *ca.* 7 μm lenses are manufactured. Dimensions, shape and roughness of these lenses are important parameters that the AFM can provide.

three-dimensional nature of AFM images can give a better idea of the shape of MEMS devices.

7.4.2.4 Optical products

Optical components can be somewhat less demanding than semiconductors, but still require very low manufacturing tolerances, and AFM can be useful to characterize several high-technology optical products. LCD screens are created by patterning a glass surface with several layers of thin films. The AFM is an excellent tool for measuring the thickness of the thin films on the surface of the glass. In the example in Figure 7.32, an image of an LCD screen is shown, with a critical measurement on a thin film step. The step height at the left is 30 nm, and the step at the right is 73 nm.

Optical components are required for many high-technology products such as optical storage device lenses, digital cameras, and fibre optics required for telecommunications. The AFM is an excellent method for imaging optical components. These components are often quite large, but measurement of these components is not possible with an optical profiler because the devices are transparent. A mechanical profiler cannot be used because it would potentially scratch their surfaces. The example in Figure 7.32 shows a micro-lens array comprising a series of microscopic lenses fabricated onto a surface. Each lens is approximately 7 microns across. The AFM is capable of precisely measuring the dimensions of each of the lenses, as well as the surface roughness of the lenses. Any anisotropy in the lenses will cause unwanted distortion in the devices that the lenses are used in.

Appendix A

AFM standards

Accurate, reproducible and quantitative image acquisition with an AFM requires some care. The most important procedure to enable reproducible and accurate results is calibration of the AFM. In order to carry out such procedures, accurate standards and references are required. Standards are used to ensure that the absolute measurements are correct, while references ensure that the instrument is giving consistent results. Most AFM calibration procedures involve using the AFM to make a measurement on a well-known sample such as the standards described in this chapter. Scanner calibration and certification procedures are described in Appendix B. Different standards for AFM calibration are available for X-Y and Z axis calibration. Apart from calibration of the scanner, the most important calibration is measurement of the tip of the probe used. The probe in AFM experiments is manufactured with rather a large tolerance, so reproducibility between probes can be quite low. However, the sharpness and aspect ratio of the probe used is one of the factors that can define the achievable resolution in an AFM experiment. Standard samples described in this chapter can enable measurement of probe radius, which greatly helps in understanding and controlling image quality. The samples and products described in this chapter are not the only ones that can be used for calibration, any sample with reproducible and well-known dimensions may be used for AFM characterization. Since any printed list of internet addresses is likely to rapidly go out of date, there is updated version of the calibration specimen supplier listing on the website accompanying this book at: http://afmhelp.com.

A1 Z axis calibration

Calibration standards and references are needed to calibrate AFMs in the vertical axis. For calibrations greater than 10 nm step height, standards or references are typically fabricated by lithographically etching patterns in a silicon or quartz substrate. It is also possible to find silicon dioxide coated with a uniform layer of metal. When calibrating the instrument for Z height measurements below 10 nm, nanoparticles with low polydispersity or atomic terraces of silicon or HOPG may be used as a reference specimen. Figure A1 shows an AFM image and measurement of a Z axis calibration specimen consisting of a pit of *ca.* 16 nm depth. Table A1 lists some Z axis calibration specimens that are commercially available.

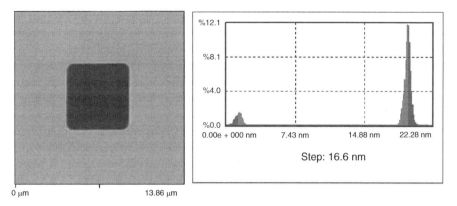

Fig. A1. AFM image is of a single etched pit in a patterned AFM standard. The histogram at the right shows the height of the feature is 16.6 nm.

As an example, Figure A2 shows an image of a common Z calibration specimen.

A2 *X-Y* Calibration and linearity

Calibration of the AFM in the *X-Y* axis so that accurate dimensional measurements can be made requires establishing the linearity and orthogonality of the AFM scanner. For large scale measurements, this requires a repeating pattern, with well controlled spacing (pitch)

Table A1. Commercial specimens for Z calibration.

Company and url	Z calibration reference
VLSI standards www.vlsistandards.com	18, 44, 100 and 180 nm (silicon) 18 nm to 8 μm (quartz)
MikroMasch www.spmtips.com	20, 100, 500 nm and 1 and 1.5 μm, HOPG
NTT AT www.ntt-at.com	Silicon monatomic steps (0.31 nm)
Ted Pella www.tedpella.com	20, 100 and 500 nm (Silicon)
Applied NanoStructures www.appnano.com	10 nm, 1 μm
Veeco www.veecoprobes.com	2, 100, or 200 nm (silicon)
NT-MDT www.ntmdt-tips.com	20, 100 or 500 nm (silicon) Silicon atomic steps (0.31 nm)
Asylum Research www.asylumresearch.com	200 nm (metal on silicon)
Nanosensors www.nanosensors.com	8 nm (silicon)

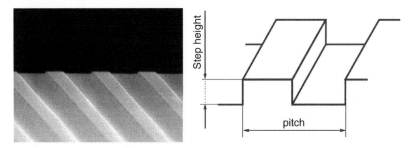

Fig. A2. An example of a Z calibration standard (TGZ sample, images courtesy of Mikromasch). These samples have a well-defined step height, enabling simple calibration of the Z axis sensitivity.

of the features. This is typically done with etched patterns in quartz or silicon. An example is shown in Figure A3. Sometimes the specimen that is used for calibrating the Z axis can be used for the X-Y axis as well. However, most specimens are accurate only in pitch *or* step height. For very high-resolution calibration, any crystalline sample that can be produced in atomic flatness and has a well-known lattice parameter could be used, but mica and HOPG are the most convenient and commonly used. Table A2 lists some available X-Y calibration specimens.

A3   *XZ* and *YZ* orthogonality

An AFM may be calibrated in X-Y and calibrated in Z and may not be useful for making measurement of angles. This is because the XZ and YZ axis may not be orthogonal. The reason for this is the crosstalk between the X-Y and Z axes as discussed in Chapters 2 and 6. With the orthogonality references, this problem can be detected, and possibly corrected for. This reference is fabricated by making a 1-D array, or line, of triangles in a silicon wafer, see Figure A4.

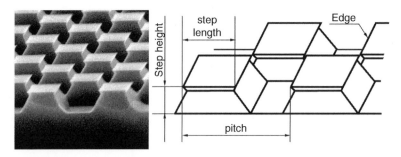

Fig. A3. Example of a silicon X-Y calibration specimen (TGX grating, Mikromasch). Images reproduced with permission from Mikromasch.

Table A2. Commercial specimens for X-Y calibration.

Company and url	X-Y calibration reference
VLSI standards www.vlsistandards.com	1.8, 3, 5, 10 and 20 μm (silicon, 2-D) 100, 200, 400, 800 and 1000 nm (silicon, 1-D)
Ted Pella www.tedpella.com	144 nm (aluminium on Silicon) 300 nm (titanium on silicon)
MikroMasch www.spmtips.com	3 and 10 μm, HOPG
SPI Supplies www.2spi.com	300 or 700 nm (metal-coated silicon)
Electron Microscopy Sciences www.emsdiasum.com	300 or 700 nm (metal-coated silicon)
Applied NanoStructures www.appnano.com	3, 10, 20 and 50 μm (metal-coated silicon)
Veeco www.veecoprobes.com	1, 2, 10, 15 μm (silicon)
NT-MDT www.ntmdt-tips.com	278 nm (aluminium on glass, 1-D) 3 μm (silicon, 2-D)
Asylum Research www.asylumresearch.com	10 and 20 μm pitch (metal on silicon)
Nanosensors www.nanosensors.com	100, 200 or 300 nm (silicon) 4, 8 and 16 μm (silicon)

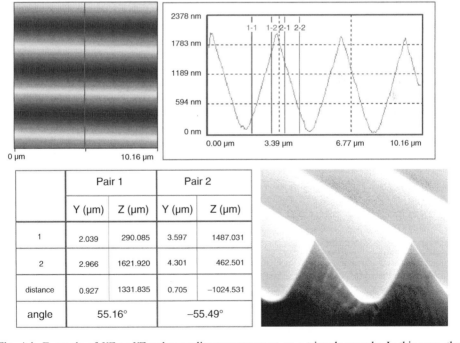

Fig. A4. Example of XZ or YZ orthogonality measurement on a triangle sample. In this case, the AFM image (top left) shows remarkably little crosstalk between the X-Y and Z axes, as the measured angles are 55.2 and 55.5° on the two sides of the triangle feature. At bottom right is an SEM image of the triangle sample (TGG sample, image courtesy of Mikromasch).

A4   Flatness

An AFM is capable of measuring surface roughness of samples at the nanometre scale. However, it is important to establish that the surface roughness measured on very flat samples is not simply the noise floor of the instrument. Also, when measuring flat samples, it is important to know the 'bow' associated with the scanner (see Chapter 2). Use of very flat reference samples is useful for establishing the performance of the instrument with respect to noise floor and bow. The use of these is described in Appendix B. The materials suitable for these sorts of measurements are atomically flat silicon wafers, and cleaved HOPG or mica. The last two are particularly convenient as cleaving them produces a very clean surface easily. These materials are available from many sources, as they have many applications outside of AFM, and just a few suppliers are given in Table A3.

A5   Particles

SPM height calibration can be done on nanoparticles of uniform size; it also verifies SPM performance on samples with weak adhesion to the surface. It can be helpful to include nanoparticles with a sample to establish the sizes of features in the images relative to the nanoparticles [279]. Larger spheres (>100 nm) tend to be polymer nano- or microparticles, whereas smaller particles are often gold nanoparticles. Spherical particles can also be used to determine the radius of the AFM probe tip [48, 49]. Some sources are given in Table A4, and examples shown in Figure A5.

Table A3. Some commercial ultra-flat specimens for out-of-plane motion and noise floor measurements.

Supplier and link	Flat substrate type
NT-MDT www.ntmdt-tips.com	HOPG, mica
SPI Supplies www.2spi.com	HOPG, mica
Novascan www.novascan.com	mica
Electron Microscopy Sciences www.emsdiasum.com	mica
Ted Pella www.tedpella.com	mica
Nanosensors www.nanosensors.com	Quartz with chromium coating

Table A4. Some commercial sources of nanoparticles for probe shape and z axis calibration.

Supplier and link	Particle type
Ted Pella www.tedpella.com	Gold colloids in 5, 15, or 15 nm diameter
Thermo Scientific www.thermo.com	Polystyrene nanospheres in a range from 20 to 900 nm
EvidentTechnology Inc. www.evidenttech.com	Quantum dots in a range from 2.2 to 5.8 nm
Electron Microscopy Sciences www.emsdiasum.com	Colloidal gold in 0.8, 6, 10, 15 and 25 nm diameters

A6 Biomolecules

Often it is necessary to verify SPM image quality on soft samples or samples with weak adhesion to the surface. It is important that this type of reference has a long-life, is stable and is indestructible. Before measuring images of 'unknown' biological molecules it is often helpful to practice on this 'known' sample. Example sources are include in Table A5.

A7 Lateral force microscopy/friction

The most well-known commercial LFM references come as 1-D arrays of triangular steps having precise linear and angular dimensions [337]. Establishing quantitative LFM data requires standardizing the AFM scanner output in the vertical and horizontal axis. Some alternatives are also available; the use of crystals with well-known angles has been

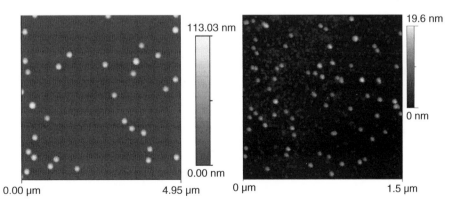

Fig. A5. Example of colloidal samples suitable for AFM calibration. Left: AFM image of 102 nm diameter polymer nanoparticles. Right: image of 15 nm gold nanoparticles. Software can be used for counting the particles and for measuring the size distribution of the particles.

Table A5. Sources of commercial biological reference samples. Right: image of NT-MDT DNA sample. Reproduced with permission from NT-MDT.

Supplier and link	Biological reference sample	
NT-MDT www.ntmdt-tips.com	Linearized plasmid DNA of known length	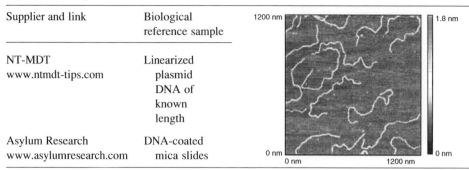
Asylum Research www.asylumresearch.com	DNA-coated mica slides	

described but these are not commercially available [77]. Another commercial possibility is an optical grating with controlled angles [735]. Some sources of commercial specimens are given in Table A6.

A8   IC-AFM mode phase imaging

In intermittent-contact mode, the AFM can measure difference in viscoelastic properties of mixed samples by phase imaging – see Section 3.2.3.2. A reference of polymer material that has hard and soft regions is used to check the AFM response for this purpose, see Table A7.

Table A6. Some commercial sources of references for lateral force calibration.

Supplier and link	LFM reference sample
Mikromasch www.spmtips.com Edmund Optics www.edmundoptics.com	Triangles (silicon), top angle 70° Steps with sloped edges (silicon), slopes 54° Ruled diffraction gratings, with various angles

Table A7. Some commercial sources of phase imaging references.

Supplier and link	Phase imaging reference sample
Electron Microscopy Sciences www.emsdiasum.com	Polymer blend sample
Asylum Research www.asylumresearch.com	SEBS polymer sample

Table A8. Some commercial sources of references to characterize tip shape.

Supplier and link	Tip check reference sample
Aurora NanoDevices www.aurorand.com	Tip check sample (100 nm z-scale)
	Nioprobe tipcheck sample (10 nm z scale)
Mikromasch www.spmtips.com	Porous aluminium
NT-MDT www.nt-mdt.com	Silicon spikes

A9   Tip shape characterization

Tip visualization helps visualize the geometry of the scanning probe *in situ*, as a conveni-ent alternative to imaging with electron microscopy. Tip radius characterization samples exhibit features sharper than an AFM tip. First-order approximation of the tip apex can be obtained. These samples can also help in troubleshooting, to identify scanning problems. A number of commercial samples are mentioned in Table A8. In addition, nanoparticles could be used (see above). Another possibility is biaxially oriented polypropylene (BOPP), which shows very fine fibres, that can only be well-resolved by a sharp probe (see Figure A6) [380, 736]. However, this material is too soft for contact-mode imaging, and is only useful for oscillating modes.

   In conclusion, making consistent, or even quantitative, measurements with an AFM requires diligence and the appropriate standard or reference. This partial list of standards and references can be used for making consistent or quantitative measurements.

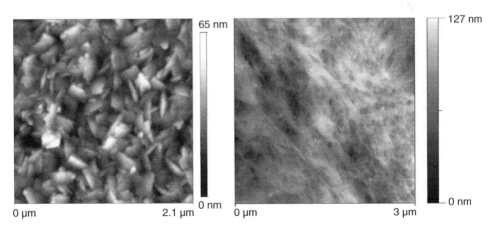

Fig. A6. AFM images of tip-checking samples. Left: 'tip check' sample from Aurora nanodevices. Right: BOPP sample, showing fine polymer fibrils.

Appendix B

Scanner calibration and certification procedures

Most AFMs operate on the principle that there's a known relation between the voltage applied to piezoelectric scanner, and the distance it moves. This relation is measured and programmed into the software (via a calibration file) at the factory where the AFM scanner is originally configured. However, the voltage–distance relation can change over time, meaning that measurements made with the AFM will be inaccurate. Thus, it is essential that a scanner certification procedure be established for any AFM. This certification should be undertaken periodically to ensure that the AFM is operating correctly and that measurements made on the AFM are meaningful. The piezoelectric scanners used in AFMs change in response as they are used. Typically a scanner will be recertified once per year. Calibration, on the other hand, implies that once the errors in the measurements made by the scanner are made, they are corrected for. Remember that if you recalibrate the scanner, the measurements before and after recalibration are not directly comparable. The procedure detailed here allows you to make the measurements required to recalibrate your scanner, but the actual procedure for changing the calibration factors varies from instrument to instrument, and should be described in the user manual. The certification procedure presented here was designed for a large (90 μm) scanner. However, this procedure can be modified to suit the needs of the instruments with other types of scanners and scan range. The different standards described in Appendix A can help with this. This procedure requires a few reference or standard samples. All of these (with the exception of the angled sample holder) are commercially available. The sources for these commercial samples were given in Appendix A. They are:

- *X-Y* test pattern: this sample has *X-Y* features that are about 1/20th the AFM full scan range. Ideally this specimen has squares on its surface. When measuring lateral dimensions the pitch (i.e. the repeat distance of the features) is the calibrated dimension.
- *Z* height standard: a sample that has a step height that is in the range of the types of application the AFM instrument is used for. For example, for measurement of features of the order of a few nanometres, a very low (e.g. 20 nm) step height could be used. For larger measurements a larger reference should be used.
- Triangle sample: etched silicon sample that has features with a known angle at its surface.

- Atomically flat sample: A highly polished sample that is contamination-free (A silicon wafer, or cleaved mica or HOPG works well).
- Angled sample holder: A small sample holder that can hold a sample at $7°$.

B1   *xy* specifications

B1.1   Scan range

An image of a standard test pattern with a known pitch is acquired. The pitch is measured from this image, ideally by averaging over a large number of cells. The scan range can be calculated by multiplying the number of cells viewed in a line profile by the pitch. See Figure B1.

B1.2   xy measurement accuracy

This test is designed to verify that *xy* measurements derived from the AFM are accurate. It also determines if they are the same across different parts of the scan range (i.e. the *x* and *y* linearity). Using the same image measured in Section B1.1, the pitch is measured at several locations on the sample. The standard deviation of the measurements is calculated, as well as the maximum error. See Figure B2.

B2   *z* specifications

B2.1   z range

Using a sample holder that applies a known angle between the sample's surface and the scanner *x-y* axis, the *xy* test pattern is scanned. The number of features in the image is counted, and from this, the *z* range of the scanner is calculated as shown in Figure B3.

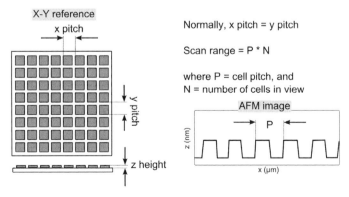

Fig. B1. Measurement of pitch with a typical *X-Y* calibration specimen.

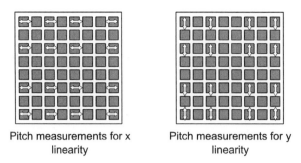

Pitch measurements for x   Pitch measurements for y
            linearity   linearity

Fig. B2. Measurements to determine x and y scan linearity.

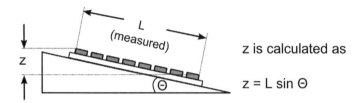

Fig. B3. Method to measure z range.

B2.2   *z calibration*

A sample that has a known z height step is scanned. A large number of such standards are available, see Appendix A. The step height is measured from the images and compared to the known dimension. There are a number of ways to measure step height. Histogram-based measurements can be used, but the image must be carefully levelled. Perhaps the most reliable method is the ISO 5436 method, using a line profile that's an average of several lines. See Chapter 5 for details on accurately measuring heights from AFM images.

B2.3   *Z linearity*

In order to determine the linearity of measurements in the z axis, the image acquired in Section B.2.1 is used. The heights of the features spread across the image, and hence across the z range are measured, and compared, as shown in Figure B4.

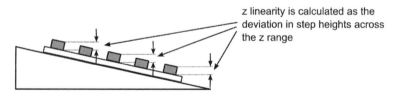

Fig. B4. Method to measure z linearity.

B2.4   z noise floor

For many applications it is essential to know the noise floor of the AFM instrument to ensure that high-resolution measurements are meaningful. This can be particularly important for high-resolution force spectroscopy. Measuring the noise floor can also help in optimizing instrument set-up and vibration isolation. It is important to know the noise floor of the z piezo in the z feedback loop as well as the noise floor of the z calibration sensor if there is one in the instrument. In most instruments, the noise floor of the z calibration sensor will be much higher than that of the z piezo.

B2.4.1   z piezo noise

This procedure is subjective, but will give a good indication of the noise floor of the AFM. In order to get reproducible results, all scan parameters should be maintained at the same values. Some factors, such as the PID values, vary greatly from instrument to instrument, so they cannot be suggested here. In each case, standard values should be established such that a fair comparison can be made.

(a) Place a flat sample in the instrument. A sample such as a silicon wafer, HOPG or mica is appropriate. The sample must be clean.
(b) Scan an image on the sample to verify cleanliness and optimize the PID parameters.
(c) Set the instrument to make a zero size scan such that the probe does not move in the x and y axis.
(d) Measure an image without probe motion in x or y, at a 1 Hz scan rate. A 256×256 pixel image is adequate. The data from the z piezo voltage should be used.
(e) Calculate the rms roughness (R_q, see Chapter 5) of the image, this value is the noise floor.

The achievable noise floor varies from one instrument to another, as well as depending on the noise in the environment, the measurement parameters, and the vibration isolation, but typically a sub-angstrom noise floor can be achieved.

B2.4.2   z calibration sensor noise

This measurement is made in exactly the same way as described above for the z piezo noise floor, with the exception that the data from the z calibration sensor is used.

B3   x-y-z coupling

Ideally motions of the scanner in the x, y and z axis would be orthogonal and independent. In practice, however, there is crosstalk between the scanning in the three axes. See Chapter 2 for the explanation of these effects.

B3.1   xy orthogonality

In order to check whether there is crosstalk between the x and y axes of scanning, the deviation from perpendicular of the features in the standard X-Y sample is measured as shown in Figure B5.

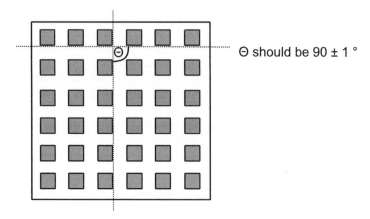

Θ should be 90 ± 1 °

Fig. B5. Method to measure orthogonality between x and y scan axes.

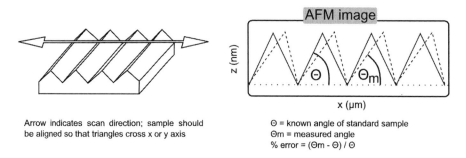

Arrow indicates scan direction; sample should be aligned so that triangles cross x or y axis

Θ = known angle of standard sample
Θm = measured angle
% error = (Θm - Θ) / Θ

Fig. B6. Procedure to measure the xz and xy crosstalk.

B3.2 xz and yz orthogonality

With the scan axis at a right angle to the ridges, an image is measured of the triangle sample (see Appendix A). The angle is measure in software. There can be a substantial deviation from the true angles due to the coupling between the x-z or y-z axes. This measurement must be done twice, in x and y axes, the sample and scan axis must be rotated between each measurement. See Figure B6.

B3.3 Out-of-plane motion

Out-of-plane motion is the sum of the non-ideal motions the AFM makes when scanning a flat sample. The major contribution to this motion is scanner bow, where applicable (see Chapter 2). In addition, x-z and y-z crosstalk will create out-of-plane motion. It is measured as shown in Figure B7. Essentially, an image of a very flat sample (such as a silicon wafer, or mica or HOPG) is measured at full scan range (or a smaller scan if

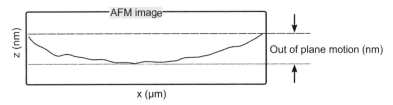

Fig. B7. Measurement of out-of-plane motion in AFM. The sample should be atomically flat, such that the out-of-plane motion measured is only due to the instrument.

required). If a full image was measured, then a line is extracted, and a first order level applied to remove sample tilt. The out-of-plane motion is measured as the full z range of the line.

Appendix C

Third party AFM software

All commercial AFM instruments are supplied with image processing display and analysis software. In general, AFM image processing and analysis is usually best done with the software supplied with the instrument. The reason for this is that this is the only software that is sure to open the AFM data acquired by the instrument correctly. Due to the very broad range of available file formats that are constantly evolving to reflect new AFM capabilities, writing and maintaining a single program to open all these formats correctly is not trivial [737]. However, there will be cases where the user requires more capabilities than are available in the supplied software. Some of the manufacturer's packages are excellent and comprehensive, but others are rather limited, and some manufacturers even go as far as to supplement them by including one of the third party packages included in this section along with their own software. If the user decides to use a third party package, it must be done with caution, and even if the files can be opened the user must check that the data is scaled correctly. This can be quickly done by measuring a few points on an unprocessed image in both the original and the third party programs and checking they are identical. The following sections describe briefly the major third party software tools available for working with AFM data. Some of these packages are commercial projects; others are academic projects. All the packages described in this chapter are in current development at the time of writing. In the accompanying website at http://afmhelp.com some other abandoned or outdated packages are also listed, as well as details of how to get all the software described. The relative merits of these packages are not reviewed here, and their order is entirely arbitrary. It is worth noting that most of these packages are aimed at SPM rather than just AFM – so they also deal with STM or SNOM files. Except where otherwise described, these packages are designed for use in MS Windows operating systems.

Gwyddion – Free SPM (AFM, SNOM/NSOM, STM, MFM, . . .) data analysis software. Freely available, open source, cross-platform software for manipulation of SPM files; this software reads many (>50) image file formats. The software is quite well documented, and as it is open source it is (in principle) possible to inspect the algorithms it uses. The package was developed by academics from the Czech metrology institute, and is updated quite regularly. Packages are available for MAC OS, Windows and Linux. All the procedures described in Chapter 5 can be carried out with Gwyddion.

Scanning Probe Image Processor – SPIP. Commercial software for manipulation of SPM files; it is probably the most comprehensive package available at the moment, but requires purchase of a license for extended use. A limited-function demonstration version is available for free download. It supports a very large number (>70) of image formats, and in some cases also opens force-curve data, and is updated quite regularly. All the procedures described in Chapter 5 can be carried out with SPIP.

Table C1. Some third party software for processing and analysis of AFM images.

Software Package	url
Gwyddion	http://www.gwyddion.net/
SPIP	http://www.imagemet.com/
WSxM	http://www.nanotec.es/
FemtoScan Online	http://www.nanoscopy.net/en/Femtoscan-D.php
PUNIAS	http://site.voila.fr/punias
Image SXM	http://www.liv.ac.uk/~Esdb/ImageSXM/
ImageJ	http://rsb.info.nih.gov/ij/

WSxM – Windows Scanning x Microscope [738]. This is commercial but free software that enables control of the AFM instruments sold by the company that developed it, but which is also a processing and analysis software package. It opens many image file formats and also some force-curve data, and it is updated regularly. All the procedures described in Chapter 5 can be carried out with WSxM.

FemtoScan Online. Commercial software from an instrument manufacturer, but which is also a processing and analysis software package that loads several other image file formats. A 30-day trial version is available for free download.

PUNIAS (Protein Unfolding and Nanoindentation Analysis Software). This free software is dedicated to analysis of force curves, and force-curve map files. It implements several of the common algorithms used for analysis of force spectroscopy and nanoindentation data. It is able to load force curve data from several of the most popular file formats. It is updated fairly regularly.

Image SXM and ImageJ. Both of these programs are versions of NIH *Image*, a public domain image processing and analysis program for the Macintosh. Image SXM is a Mac OS-only version of NIH *Image* that has been extended to handle the loading, display and analysis of AFM images in several file formats. Image J is a cross-platform image analysis program, again based on NIH *Image*. Packages are available for Mac OS, Linux and Windows. It is not specifically designed for SPM images, but there are plugins available to load a few file formats. These programs are rather limited in their applicability to AFM files, as they are not by default designed to work with three-dimensional topographical data, but have many more 'classical' image analysis procedures, and so can be useful under some circumstances.

A list of urls linking to all the software described here is reproduced in Table C1. An updated listing is available at http://afmhelp.com.

Bibliography

[1] Hoffmann, R.; Baratoff, A.; Hug, H. J.; Hidber, H. R.; von Lohneysen, H.; Guntherodt, H. J. G., Mechanical manifestations of rare atomic jumps in dynamic force microscopy. *Nanotechnology* 2007, 18 (39), 395503.

[2] Crampton, N.; Yokokawa, M.; Dryden, D. T. F.; Edwardson, J. M.; Rao, D. N.; Takeyasu, K.; Yoshimura, S. H.; Henderson, R. M., Fast-scan atomic force microscopy reveals that the type III restriction enzyme EcoP15I is capable of DNA translocation and looping. *Proceedings of the National Academy of Sciences of the United States of America* 2007, 104 (31), 12755–60.

[3] Ando, T.; Uchihashi, T.; Kodera, N.; Yamamoto, D.; Taniguchi, M.; Miyagi, A.; Yamashita, H., High-speed atomic force microscopy for observing dynamic biomolecular processes. *Journal of Molecular Recognition* 2007, 20 (6), 448–58.

[4] Yokokawa, M.; Yoshimura, S. H.; Naito, Y.; Ando, T.; Yagi, A.; Sakai, N.; Takeyasu, K., Fast-scanning atomic force microscopy reveals the molecular mechanism of DNA cleavage by ApaI endonuclease. *IEE Proceedings – Nanobiotechnology* 2006, 153 (4), 60–66.

[5] Sullivan, C. J.; Morrell, J. L.; Allison, D. P.; Doktycz, M. J., Mounting of *Escherichia coli* spheroplasts for AFM imaging. *Ultramicroscopy* 2005, 105 (1–4), 96–102.

[6] Doktycz, M. J.; Sullivan, C. J.; Hoyt, P. R.; Pelletier, D. A.; Wu, S.; Allison, D. P., AFM imaging of bacteria in liquid media immobilized on gelatin coated mica surfaces. *Ultramicroscopy* 2003, 97 (1–4), 209–16.

[7] Schimmel, T.; Koch, T.; Kuppers, J.; Lux-Steiner, M., True atomic resolution under ambient conditions obtained by atomic force microscopy in the contact mode. *Applied Physics A: Materials Science & Processing* 1999, 68 (4), 399–402.

[8] Sugimoto, Y.; Pou, P.; Abe, M.; Jelinek, P.; Perez, R.; Morita, S.; Custance, O., Chemical identification of individual surface atoms by atomic force microscopy. *Nature* 2007, 446 (7131), 64–67.

[9] Tromas, C.; Eaton, P.; Mimault, J.; Rojo, J.; Penadés, S., Structural characterization of self-assembled monolayers of neoglycoconjugates using atomic force microscopy. *Langmuir* 2005, 21 (14), 6142–44.

[10] Shmalz, G., Uber Glatte und Ebenheit als physikalisches und physiologishes Problem. *Verein Deutscher Ingenieure* 1929, 1461–67.

[11] Becker, H.; Bender, O.; Bergmann, L.; Rost, K.; Zobel, A. Apparatus for measuring surface irregularities. United States Patent number: 2728222, 1955.

[12] Young, R.; Ward, J.; Scire, F., The topografiner: an instrument for measuring surface micro-topography. *Review of Scientific Instruments* 1972, 43 (7), 999–1011.

[13] Binnig, G.; Rohrer, H.; Gerber, C.; Weibel, E., Surface studies by scanning tunneling microscopy. *Physical Review Letters* 1982, 49 (1), 57–61.

[14] Binnig, G.; Rohrer, H., Scanning tunneling microscopy. *Helvetica Physica Acta* 1982, 55 (6), 726–35.

[15] Binnig, G.; Rohrer, H., Scanning tunneling microscopy. *Surface Science* 1983, 126 (1–3), 236–44.

[16] Binnig, G.; Rohrer, H.; Gerber, C.; Weibel, E., 7×7 reconstruction on Si(111) resolved in real space. *Physical Review Letters* 1983, 50 (2), 120–23.

[17] The Nobel Prize in Physics 1986. http://nobelprize.org/nobel_prizes/physics/laureates/1986/

[18] Feenstra, R. M.; Lutz, M. A. *Scanning tunneling microscopy and spectroscopy of the Si(111) 5 × 5 surface*, Fifth international conference on scanning tunneling microscopy/spectroscopy. AVS: Boston, MA, 1991; pp 716–20.

[19] Binnig, G.; Quate, C. F.; Gerber, C., Atomic force microscope. *Physical Review Letters* 1986, 56 (9), 930–33.

[20] Binnig, G. Atomic force microscope and method for imaging surfaces with atomic resolution. United States Patent number: 4724318, 02/09/1988, 1998.

[21] Binnig, G.; Gerber, C.; Stoll, E.; Albrecht, T. R.; Quate, C. F., Atomic resolution with atomic force microscope. *Europhysics Letters* 1987, 3 (12), 1281–86.

[22] Albrecht, T. R.; Akamine, S.; Carver, T. E.; Quate, C. F., Microfabrication of cantilever styli for the atomic force microscope. *Journal of Vacuum Science & Technology A – Vacuum Surfaces and Films* 1990, 8 (4), 3386–96.

[23] Meyer, G.; Amer, N. M., Novel optical approach to atomic force microscopy. *Applied Physics Letters* 1988, 53 (12), 1045–47.

[24] Alexander, S.; Hellemans, L.; Marti, O.; Schneir, J.; Elings, V.; Hansma, P. K.; Longmire, M.; Gurley, J., An atomic-resolution atomic-force microscope implemented using an optical lever. *Journal of Applied Physics* 1989, 65 (1), 164–67.

[25] Chen, C. J., *Introduction to Scanning Tunneling Microscopy*, second edition. Oxford University Press: Oxford, 2007.

[26] Bonnell, D. A., Scanning tunneling microscopy and spectroscopy of oxide surfaces. *Progress in Surface Science* 1998, 57 (3), 187–252.

[27] Kwon, J.; Hong, J.; Kim, Y. S.; Lee, D. Y.; Lee, K.; Lee, S. M.; Park, S. I., Atomic force microscope with improved scan accuracy, scan speed, and optical vision. *Review of Scientific Instruments* 2003, 74 (10), 4378–83.

[28] Dixson, R.; Koning, R.; Vorburger, T. V.; Fu, J.; Tsai, V. W., Measurement of pitch and width samples with the NIST calibrated atomic force microscope. In *Metrology, Inspection, and Process Control for Microlithography Xii*, Singh, B., Ed. 1998; Vol. 3332, pp 420–32.

[29] Youm, W.; Jung, J.; Lee, S.; Park, K., Control of voice coil motor nanoscanners for an atomic force microscopy system using a loop shaping technique. *Review of Scientific Instruments* 2008, 79 (1), 013707–6.

[30] Xie, H.; Rakotondrabe, M.; Regnier, S., Characterizing piezoscanner hysteresis and creep using optical levers and a reference nanopositioning stage. *Review of Scientific Instruments* 2009, 80 (4), 046102–3.

[31] Barrett, R. C.; Quate, C. F., Optical scan-correction system applied to atomic force microscopy. *Review of Scientific Instruments* 1991, 62 (6), 1393–99.

[32] Cronin, P. J.; Fekete, P. W.; Arnison, M. R.; Cogswell, C. J., Characterization of an open-loop controlled scanning stage using a knife edge optical technique. *Review of Scientific Instruments* 2000, 71 (1), 118–23.

[33] Huang, Q.; Gonda, S.; Misumi, I.; Sato, O.; Keem, T.; Kurosawa, T., Nonlinear and hysteretic influence of piezoelectric actuators in AFMs on lateral dimension measurement *Sensors and Actuators A: Physical* 2006, 125 (2), 590–96.

[34] Ando, T.; Kodera, N.; Takai, E.; Maruyama, D.; Saito, K.; Toda, A., A high-speed atomic force microscope for studying biological macromolecules. *Proceedings of the National Academy of Sciences of the United States of America* 2001, 98 (22), 12468–72.

[35] Picco, L. M.; Bozec, L.; Ulcinas, A.; Engledew, D. J.; Antognozzi, M.; Horton, M. A.; Miles, M. J., Breaking the speed limit with atomic force microscopy. *Nanotechnology* 2007, 18 (4), 044030.

[36] Humphris, A. D. L.; Miles, M. J.; Hobbs, J. K., A mechanical microscope: high-speed atomic force microscopy. *Applied Physics Letters* 2005, 86 (3), 034106.

[37] Ruf, A.; Abraham, M.; Lacher, M.; Mayr, K.; Zetterer, T. *A miniaturised Fabry Perot AFM sensor*, The 8th International Conference on Solid-State Sensors and Actuators, 1995; pp 660–63.

[38] Peng, Z.; West, P., Crystal sensor for microscopy applications. *Applied Physics Letters* 2005, 86 (1), 014107–3.

[39] Kim, M. S.; Choi, J. H.; Park, Y. K.; Kim, J. H., Atomic force microscope cantilever calibration device for quantified force metrology at micro- or nano-scale regime: the nano force calibrator (NFC). *Metrologia* 2006, 43 (5), 389–95.

[40] Tortonese, M.; Yamada, H.; Barrett, R. C.; Quate, C. F. *Atomic force microscopy using a piezoresistive cantilever*, TRANSDUCERS '91. International Conference on Solid-State Sensors and Actuators. Digest of Technical Papers, San Francisco, CA, 1991; pp 448–51.

[41] Putman, C. A. J.; Degrooth, B. G.; Vanhulst, N. F.; Greve, J., A theoretical comparison between interferometric and optical beam deflection technique for the measurement of cantilever displacement in AFM. *Ultramicroscopy* 1992, 42, 1509–13.

[42] Neagu, C.; van der Werf, K. O.; Putman, C. A. J.; Kraan, Y. M.; de Grooth, B. G.; van Hulst, N. F.; Greve, J., Analysis of immunolabeled cells by atomic force microscopy, optical microscopy, and flow cytometry. *Journal of Structural Biology* 1994, 112 (1), 32–40.

[43] Nakano, K., A novel low profile atomic force microscope compatible with optical microscopes. *Review of Scientific Instruments* 1998, 69 (3), 1406.

[44] Clifford, C. A.; Seah, M. P., The determination of atomic force microscope cantilever spring constants via dimensional methods for nanomechanical analysis. *Nanotechnology* 2005, 16 (9), 1666.

[45] Vick, D.; Brett, M. J.; Westra, K., Porous thin films for the characterization of atomic force microscope tip morphology. *Thin Solid Films* 2002, 408 (1–2), 79–86.

[46] DeRose, J. A.; Revel, J. P., Examination of atomic (scanning) force microscopy probe tips with the transmission electron microscope. *Microscopy and Microanalysis* 1997, 3 (3), 203–13.

[47] Chung, K. H.; Kim, D. E., Wear characteristics of diamond-coated atomic force microscope probe. *Ultramicroscopy* 2007, 108 (1), 1–10.

[48] Ramirez-Aguilar, K. A.; Rowlen, K. L., Tip characterization from AFM images of nanometric spherical particles. *Langmuir* 1998, 14 (9), 2562–66.

[49] Zeng, Z.-g.; Zhu, G.-d.; Guo, Z.; Zhang, L.; Yan, X.-j.; Du, Q.-g.; Liu, R., A simple method for AFM tip characterization by polystyrene spheres. *Ultramicroscopy* 2008, 108 (9), 975–80.

[50] Villarrubia, J. S., Scanned probe microscope tip characterization without calibrated tip characterizers. *Journal of Vacuum Science & Technology B* 1996, 14 (2), 1518–21.

[51] Williams, P. M.; Shakesheff, K. M.; Davies, M. C.; Jackson, D. E.; Roberts, C. J.; Tendler, S. J. B., Blind reconstruction of scanning probe image data. *Journal of Vacuum Science & Technology B* 1996, 14 (1557), 1557–62.

[52] Villarrubia, J. S., Morphological estimation of tip geometry for scanned probe microscopy. *Surface Science* 1994, 321 (3), 287–300.

[53] Villarrubia, J. S., Algorithms for scanned probe microscope image simulation, surface reconstruction, and tip estimation. *Journal of Research of the National Institute of Standards and Technology* 1997, 102 (4), 425–54.

[54] Williams, P. M.; Shakesheff, K. M.; Davies, M. C.; Jackson, D. E.; Roberts, C. J.; Tendler, S. J. B., Toward true surface recovery: studying distortions in scanning probe microscopy image data. *Langmuir* 1996, 12 (14), 3468–71.

[55] Dongmo, L. S.; Villarrubia, J. S.; Jones, S. N.; Renegar, T. B.; Postek, M.; Song, J. F., Experimental test of blind tip reconstruction for scanning probe microscopy. *Ultramicroscopy* 2000, 85 (3), 141–53.

[56] Todd, B. A.; Eppell, S. J., A method to improve the quantitative analysis of SFM images at the nanoscale. *Surface Science* 2001, 491 (3), 473–83.

[57] Tranchida, D.; Piccarolo, S.; Deblieck, R. A. C., Some experimental issues of AFM tip blind estimation: the effect of noise and resolution. *Measurement Science and Technology* 2006, 10, 2630–36.

[58] Kitching, S.; Williams, P. M.; Roberts, C. J.; Davies, M. C.; Tendler, S. J. B., Quantifying surface topography and scanning probe image reconstruction. *Journal of Vacuum Science & Technology B* 1999, 17 (2), 273–79.

[59] Emerson, R. J.; Camesano, T. A., On the importance of precise calibration techniques for an atomic force microscope. *Ultramicroscopy* 2006, 106 (4–5), 413–22.

[60] Higgins, M. J.; Proksch, R.; Sader, J. E.; Polcik, M.; Mc Endoo, S.; Cleveland, J. P.; Jarvis, S. P., Noninvasive determination of optical lever sensitivity in atomic force microscopy. *Review of Scientific Instruments* 2006, 77 (1), 013701.

[61] Green, C. P.; Lioe, H.; Cleveland, J. P.; Proksch, R.; Mulvaney, P.; Sader, J. E., Normal and torsional spring constants of atomic force microscope cantilevers. *Review of Scientific Instruments* 2004, 75 (6), 1988–96.

[62] Matei, G. A.; Thoreson, E. J.; Pratt, J. R.; Newell, D. B.; Burnham, N. A., Precision and accuracy of thermal calibration of atomic force microscopy cantilevers. *Review of Scientific Instruments* 2006, 77 (8), 083703.

[63] Burnham, N. A.; Chen, X.; Hodges, C. S.; Matei, G. A.; Thoreson, E. J.; Roberts, C. J.; Davies, M. C.; Tendler, S. J. B., Comparison of calibration methods for atomic-force microscopy cantilevers. *Nanotechnology* 2003, 14 (1), 1–6.

[64] Gibson, C. T.; Smith, D. A.; Roberts, C. J., Calibration of silicon atomic force microscope cantilevers. *Nanotechnology* 2005, 16 (2), 234–38.

[65] Gibson, C. T.; Watson, G. S.; Myhra, S., Scanning force microscopy – Calibrative procedures for 'best practice'. *Scanning* 1997, 19 (8), 564–81.

[66] Sader, J. E.; Chon, J. W. M.; Mulvaney, P., Calibration of rectangular atomic force microscope cantilevers. *Review of Scientific Instruments* 1999, 70 (10), 3967–69.

[67] Sader, J. E.; Larson, I.; Mulvaney, P.; White, L. R., Method for the calibration of atomic force microscope cantilevers. *Review of Scientific Instruments* 1995, 66 (7), 3789–98.

[68] Sader, J. E., Frequency response of cantilever beams immersed in viscous fluids with applications to the atomic force microscope. *Journal of Applied Physics* 1998, 84 (1), 64–76.

[69] Sader, J. E.; Pacifico, J.; Green, C. P.; Mulvaney, P., General scaling law for stiffness measurement of small bodies with applications to the atomic force microscope. *Journal of Applied Physics* 2005, 97 (12), 124903–7.

[70] Cleveland, J. P.; Manne, S.; Bocek, D.; Hansma, P. K., A nondestructive method for determining the spring constant of cantilevers for scanning force microscopy. *Review of Scientific Instruments* 1993, 64 (2), 403–5.

[71] Jing, G. Y.; Ma, J.; Yu, D. P., Calibration of the spring constant of AFM cantilever. *Journal of Electron Microscopy* 2007, 56 (1), 21–25.

[72] Hutter, J. L.; Bechhoefer, J., Calibration of atomic-force microscope tips. *Review of Scientific Instruments* 1993, 64 (7), 1868–73.

[73] Hutter, J. L., Comment on tilt of atomic force microscope cantilevers: effect on spring constant and adhesion measurements. *Langmuir* 2005, 21 (6), 2630–32.

[74] Gates, R. S.; Reitsma, M. G., Precise atomic force microscope cantilever spring constant calibration using a reference cantilever array. *Review of Scientific Instruments* 2007, 78 (8), 086101–3.

[75] Tortonese, M.; Kirk, M., Characterization of application specific probes for SPMs. In *Micromachining and Imaging*, Michalske, T. A.; Wendman, M. A., Eds. SPIE – Int. Soc. Optical Engineering: Bellingham, 1997; Vol. 3009, pp 53–60.

[76] Bogdanovic, G.; Meurk, A.; Rutland, M. W., Tip friction – torsional spring constant determination. *Colloids and Surfaces B: Biointerfaces* 2000, 19 (4), 397-405.

[77] Ogletree, D. F.; Carpick, R. W.; Salmeron, M., Calibration of frictional forces in atomic force microscopy. *Review of Scientific Instruments* 1996, 67 (9), 3298-3306.

[78] Minne, S. C.; Yaralioglu, G.; Manalis, S. R.; Adams, J. D.; Zesch, J.; Atalar, A.; Quate, C. F., Automated parallel high-speed atomic force microscopy. *Applied Physics Letters* 1998, 72 (18), 2340-42.

[79] Viani, M. B.; Schaffer, T. E.; Paloczi, G. T.; Pietrasanta, L. I.; Smith, B. L.; Thompson, J. B.; Richter, M.; Rief, M.; Gaub, H. E.; Plaxco, K. W.; Cleland, A. N.; Hansma, H. G.; Hansma, P. K., Fast imaging and fast force spectroscopy of single biopolymers with a new atomic force microscope designed for small cantilevers. *Review of Scientific Instruments* 1999, 70 (11), 4300-3.

[80] Hansma, P. K.; Schitter, G.; Fantner, G. E.; Prater, C., Applied physics – high-speed atomic force microscopy. *Science* 2006, 314 (5799), 601-2.

[81] Fantner, G. E.; Schitter, G.; Kindt, J. H.; Ivanov, T.; Ivanova, K.; Patel, R.; Holten-Andersen, N.; Adams, J.; Thurner, P. J.; Rangelow, I. W.; Hansma, P. K., Components for high speed atomic force microscopy. *Ultramicroscopy* 2006, 106 (8–9), 881–87.

[82] Viani, M. B.; Schaffer, T. E.; Chand, A.; Rief, M.; Gaub, H. E.; Hansma, P. K., Small cantilevers for force spectroscopy of single molecules. *Journal of Applied Physics* 1999, 86 (4), 2258-62.

[83] Walters, D. A.; Cleveland, J. P.; Thomson, N. H.; Hansma, P. K.; Wendman, M. A.; Gurley, G.; Elings, V., Short cantilevers for atomic force microscopy. *Review of Scientific Instruments* 1996, 67 (10), 3583-90.

[84] Schäffer, T. E.; Hansma, P. K., Characterization and optimization of the detection sensitivity of an atomic force microscope for small cantilevers. *Journal of Applied Physics* 1998, 84 (9), 4661-66.

[85] Burnham, N. A.; Colton, H. M.; Pollock, H. M., Interpretation of force curves in force microscopy. *Nature* 1993, 4, 64-80.

[86] Martin, C.; Murano, F. P.; Dagata, J. A., Measurements of electrical conductivity of a nanometer-scale water meniscus by atomic force microscopy. *2003 Third IEEE Conference on Nanotechnology, Vols One and Two, Proceedings* 2003, 781-84.

[87] Lee, J.; Chae, J.; Kim, C. K.; Kim, H.; Oh, S.; Kuk, Y., Versatile low-temperature atomic force microscope with *in situ* piezomotor controls, charge-coupled device vision, and tip-gated transport measurement capability. *Review of Scientific Instruments* 2005, 76 (9), 093701-5.

[88] Broekmaat, J.; Brinkman, A.; Blank, D. H. A.; Rijnders, G., High temperature surface imaging using atomic force microscopy. *Applied Physics Letters* 2008, 92 (4), 043102-3.

[89] Shao, Z. F.; Zhang, Y. Y., Biological cryo atomic force microscopy: a brief review. *Ultramicroscopy* 1996, 66 (3-4), 141-52.

[90] Han, W. H.; Mou, J. X.; Sheng, J.; Yang, J.; Shao, Z. F., Cryo atomic force microscopy: a new approach for biological imaging at high resolution. *Biochemistry* 1995, 34 (26), 8215-20.

[91] Sheng, S.; Czajkowsky, D. M.; Shao, Z., Localization of linker histone in chromatosomes by cryo-atomic force microscopy. *Biophysical Journal* 2006, 91 (4), L35-37.

[92] Wu, J. J.; Reading, M.; Craig, D. Q. A., Application of calorimetry, sub-ambient atomic force microscopy and dynamic mechanical analysis to the study of frozen aqueous trehalose solutions. *Pharmaceutical Research* 2008, 25 (6), 1396-1404.

[93] Hug, H. J.; Stiefel, B.; van Schendel, P. J. A.; Moser, A.; Martin, S.; Guntherodt, H. J., A low temperature ultrahigh vaccum scanning force microscope. *Review of Scientific Instruments* 1999, 70 (9), 3625-40.

[94] Friedbacher, G.; Fuchs, H., Classification of scanning probe microscopies (Technical report). *Pure and Applied Chemistry* 1999, 71 (7), 1337–57.

[95] Hecht, B.; Sick, B.; Wild, U. P.; Deckert, V.; Zenobi, R.; Martin, O. J. F.; Pohl, D. W., Scanning near-field optical microscopy with aperture probes: fundamentals and applications. *Journal of Chemical Physics* 2000, 112 (18), 7761–74.

[96] Rasmussen, A.; Deckert, V., New dimension in nano-imaging: breaking through the diffraction limit with scanning near-field optical microscopy. *Analytical and Bioanalytical Chemistry* 2005, 381 (1), 165–72.

[97] Zhong, Q.; Inniss, D.; Kjoller, K.; Elings, V. B., Fractured polymer/silica fiber surface studied by tapping mode atomic force microscopy. *Surface Science* 1993, 290 (1–2), L688–L692.

[98] Karrasch, S.; Dolder, M.; Schabert, F.; Ramsden, J.; Engel, A., Covalent binding of biological samples to solid supports for scanning probe microscopy in buffer solution. *Biophysical Journal* 1993, 65 (6), 2437–46.

[99] Mou, J. X.; Czajkowsky, D. M.; Sheng, S. J.; Ho, R. Y.; Shao, Z. F., High resolution surface structure of E-coli GroES oligomer by atomic force microscopy. *FEBS Letters* 1996, 381 (1–2), 161–64.

[100] Gonçalves, R. P.; Busselez, J.; Lévy, D.; Seguin, J.; Scheuring, S., Membrane insertion of Rhodopseudomonas acidophila light harvesting complex 2 investigated by high resolution AFM. *Journal of Structural Biology* 2005, 149 (1), 79–86.

[101] Le Grimellec, C.; Lesniewska, E.; Giocondi, M.-C.; Finot, E.; Vie, V.; Goudonnet, J.-P., Imaging of the surface of living cells by low-force contact-mode atomic force microscopy. *Biophysical Journal* 1998, 75 (2), 695–703.

[102] Müller, D. J.; Schoenenberger, C. A.; Schabert, F.; Engel, A., Structural changes in native membrane proteins monitored at subnanometer resolution with the atomic force microscope: a review. *Journal of Structural Biology* 1997, 119 (2), 149–57.

[103] Murphy, M. F.; Lalor, M. J.; Manning, F. C. R.; Lilley, F.; Crosby, S. R.; Randall, C.; Burton, D. R., Comparative study of the conditions required to image live human epithelial and fibroblast cells using atomic force microscopy. *Microscopy Research and Technique* 2006, 69 (9), 757–65.

[104] Möller, C.; Allen, M.; Elings, V.; Engel, A.; Müller, D. J., Tapping-mode atomic force microscopy produces faithful high-resolution images of protein surfaces. *Biophysical Journal* 1999, 77 (2), 1150–58.

[105] Martin, Y.; Wickramasinghe, H. K., Magnetic imaging by force microscopy with 1000-Å resolution. *Applied Physics Letters* 1987, 50 (20), 1455–57.

[106] Morita, S.; Giessibl, F. J.; Sugawara, Y.; Hosoi, H.; Mukasa, K.; Sasahara, A.; Onishi, H., Noncontact atomic force microscopy and its related topics. In *Nanotribology and Nanomechanics*, 2005; pp 141–83.

[107] Morita, S.; Wiesendanger, R.; Meyer, E., *Noncontact Atomic Force Microscopy*. Springer: 2002.

[108] García, R.; Pérez, R., Dynamic atomic force microscopy methods. *Surface Science Reports* 2002, 47 (6–8), 197–301.

[109] Martin, Y.; Williams, C. C.; Wickramasinghe, H. K., Atomic force microscope–force mapping and profiling on a sub 100 Å scale. *Journal of Applied Physics* 1987, 61 (10), 4723–29.

[110] Ho, H.; West, P., Optimizing AC-mode atomic force microscope imaging. *Scanning* 1996, 18 (5), 339–43.

[111] Giessibl, F. J.; Bielefeldt, H.; Hembacher, S.; Mannhart, J., Calculation of the optimal imaging parameters for frequency modulation atomic force microscopy. *Applied Surface Science* 1999, 140 (3–4), 352–57.

[112] Herminghaus, S.; Fery, A.; Reim, D., Imaging of droplets of aqueous solutions by tapping-mode scanning force microscopy. *Ultramicroscopy* 1997, 69 (3), 211–17.

[113] Checco, A.; Schollmeyer, H.; Daillant, J.; Guenoun, P.; Boukherroub, R., Nanoscale wettability of self-assembled monolayers investigated by noncontact atomic force microscopy. *Langmuir* 2006, 22 (1), 116–26.

[114] Yang, C.-W.; Hwang, I.-S.; Chen, Y. F.; Chang, C. S.; Tsai, D. P., Imaging of soft matter with tapping-mode atomic force microscopy and non-contact-mode atomic force microscopy. *Nanotechnology* 2007, 18 (8), 084009.

[115] Hoogenboom, B. W.; Hug, H. J.; Pellmont, Y.; Martin, S.; Frederix, P. L. T. M.; Fotiadis, D.; Engel, A., Quantitative dynamic-mode scanning force microscopy in liquid. *Applied Physics Letters* 2006, 88 (19), 193109–3.

[116] Zimmermann, H.; Hagedorn, R.; Richter, E.; Fuhr, G., Topography of cell traces studied by atomic force microscopy. *European Biophysics Journal* 1999, 28 (6), 516–25.

[117] Albrecht, T. R.; Grutter, P.; Horne, D.; Rugar, D., Frequency modulation detection using high-Q cantilevers for enhanced force microscope sensitivity. *Journal of Applied Physics* 1991, 69 (2), 668–73.

[118] Sugawara, Y.; Ohta, M.; Ueyama, H.; Morita, S., Defect motion on an InP(110) surface observed with noncontact atomic force microscopy. *Science* 1995, 270 (5242), 1646–48.

[119] Luthi, R.; Meyer, E.; Bammerlin, M.; Baratoff, A.; Lehmann, T.; Howald, L.; Gerber, C.; Guntherodt, H. J., Atomic resolution in dynamic force microscopy across steps on Si(111) 7×7. *Zeitschrift Fur Physik B – Condensed Matter* 1996, 100 (2), 165–67.

[120] Sugimoto, Y.; Abe, M.; Konoshita, S.; Morita, S., Direct observation of the vacancy site of the iron silicide c(4 × 8) phase using frequency modulation atomic force microscopy. *Nanotechnology* 2007, 18 (8), 084012.

[121] Seino, Y.; Yoshikawa, S.; Abe, M.; Morita, S., Growth dynamics of insulating SrF2 films on Si(111). *Journal of Physics – Condensed Matter* 2007, 19 (44), 9.

[122] Sugimoto, Y.; Abe, M.; Hirayama, S.; Morita, S., Highly resolved non-contact atomic force microscopy images of the Sn/Si(111)-(2 root 3 × 2 root 3) surface. *Nanotechnology* 2006, 17 (16), 4235–39.

[123] Eaton, P.; Ragusa, A.; Clavel, C.; Rojas, C. T.; Graham, P.; Duran, R. V.; Penades, S., Glyconanoparticle–DNA interactions: an atomic force microscopy study. *IEEE Transactions on Nanobioscience* 2007, 6 (4), 309–18.

[124] Han, W.; Lindsay, S. M.; Jing, T., A magnetically driven oscillating probe microscope for operation in liquids. *Applied Physics Letters* 1996, 69 (26), 4111–13.

[125] Revenko, I.; Proksch, R., Magnetic and acoustic tapping mode microscopy of liquid phase phospholipid bilayers and DNA molecules. *Journal of Applied Physics* 2000, 87 (1), 526–33.

[126] Putman, C. A. J.; Van der Werf, K. O.; De Grooth, B. G.; Van Hulst, N. F.; Greve, J., Tapping mode atomic force microscopy in liquid. *Applied Physics Letters* 1994, 64 (18), 2454–56.

[127] Volkov, A. O.; Burnell-Gray, J. S.; Datta, P. K., Frequency response of atomic force microscope cantilever driven by fluid. *Applied Physics Letters* 2004, 85 (22), 5397–99.

[128] Schäffer, T. E.; Cleveland, J. P.; Ohnesorge, F.; Walters, D. A.; Hansma, P. K., Studies of vibrating atomic force microscope cantilevers in liquid. *Journal of Applied Physics* 1996, 80 (7), 3622–7.

[129] Connell, S. D.; Smith, D. A., The atomic force microscope as a tool for studying phase separation in lipid membranes (review). *Molecular Membrane Biology* 2006, 23 (1), 17–28.

[130] Lantz, M.; Liu, Y. Z.; Cui, X. D.; Tokumoto, H.; Lindsay, S. M. *Dynamic force microscopy in fluid*, 3rd Conference on Development and Industrial Application of Scanning Probe Methods (SXM-3), Switzerland, John Wiley, 1998; pp 354–60.

[131] Rodriguez, T. R.; García, R., Compositional mapping of surfaces in atomic force microscopy by excitation of the second normal mode of the microcantilever. *Applied Physics Letters* 2004, 84 (3), 449–51.

[132] Stark, R. W.; Drobek, T.; Heckl, W. M., Tapping-mode atomic force microscopy and phase-imaging in higher eigenmodes. *Applied Physics Letters* 1999, 74 (22), 3296–98.

[133] Martínez, N. F.; Lozano, J. R.; Herruzo, E. T.; Garcia, F.; Richter, C.; Sulzbach, T.; García, R., Bimodal atomic force microscopy imaging of isolated antibodies in air and liquids. *Nanotechnology* 2008, 19 (38), 384011.

[134] Lozano, J. R.; García, R., Theory of multifrequency atomic force microscopy. *Physical Review Letters* 2008, 1 (7), 076102.

[135] Crittenden, S.; Raman, A.; Reifenberger, R., Probing attractive forces at the nanoscale using higher-harmonic dynamic force microscopy. *Physical Review B* 2005, 72 (23), 235422.

[136] Proksch, R., Multifrequency, repulsive-mode amplitude-modulated atomic force microscopy. *Applied Physics Letters* 2006, 89 (11), 113121–3.

[137] Stark, R. W.; Heckl, W. M., Higher harmonics imaging in tapping-mode atomic-force microscopy. *Review of Scientific Instruments* 2003, 74 (12), 5111–14.

[138] Patil, S.; Martinez, N. F.; Lozano, J. R.; García, R., Force microscopy imaging of individual protein molecules with sub-pico newton force sensitivity. *Journal of Molecular Recognition* 2007, 20, 516–23.

[139] Giessibl, F. J., Higher-harmonic atomic force microscopy. *Surface and Interface Analysis* 2006, 38 (12–13), 1696–1701.

[140] Stroscio, J. A.; Feenstra, R. M.; Fein, A. P., Electronic structure of the Si(111)2 × 1 surface by scanning-tunneling microscopy. *Physical Review Letters* 1986, 57 (20), 2579.

[141] Chen, C. J., Scanning tunneling spectroscopy. In *Introduction to Scanning Tunneling Microscopy*, second edition. Oxford University Press: Oxford, 2008; pp 331–48.

[142] Butt, H.-J.; Cappella, B.; Kappl, M., Force measurements with the atomic force microscope: technique, interpretation and applications. *Surface Science Reports* 2005, 59 (1–6), 1–152.

[143] Hinterdorfer, P.; Baumgartner, W.; Gruber, H. J.; Schilcher, K.; Schindler, H., Detection and localization of individual antibody-antigen recognition events by atomic force microscopy. *Proceedings of the National Academy of Sciences of the United States of America* 1996, 93 (8), 3477–81.

[144] Tromas, C.; Rojo, J.; de la Fuente, J. M.; Barrientos, A. G.; García, R.; Penadés, S., Adhesion forces between Lewis X determinant antigens as measured by atomic force microscopy. *Angewandte Chemie International Edition* 2001, 40 (16), 3052–5.

[145] Clausen-Schaumann, H.; Seitz, M.; Krautbauer, R.; Gaub, H. E., Force spectroscopy with single bio-molecules. *Current Opinion in Chemical Biology* 2000, 4 (5), 524–30.

[146] Eaton, P.; Graham, P.; Smith, J. R.; Smart, J. D.; Nevell, T. G.; Tsibouklis, J., Mapping the surface heterogeneity of a polymer blend: an adhesion-force-distribution study using the atomic force microscope. *Langmuir* 2000, 16 (21), 7887–90.

[147] Subramanian, S.; Sampath, S., Effect of chain length on the adhesion behaviour of n-alkanethiol self-assembled monolayers on Au(111): an atomic force microscopy study. *Pramana – Journal of Physics* 2005, 65 (4), 753–61.

[148] Ralston, J.; Larson, I.; Rutland, M. W.; Feiler, A. A.; Kleijn, M., Atomic force microscopy and direct surface force measurements (IUPAC technical report). *Pure and Applied Chemistry* 2005, 77 (12), 2149–70.

[149] Ducker, W. A.; Senden, T. J.; Pashley, R. M., Direct measurement of colloidal forces using an atomic force microscope. *Nature* 1991, 353 (6341), 239–41.

[150] Butt, H.-J.; Jaschke, M.; Ducker, W., Measuring surface forces in aqueous electrolyte solution with the atomic force microscope. *Bioelectrochemistry and Bioenergetics* 1995, 38 (1), 191–201.

[151] Emerson, R. J.; Camesano, T. A., Nanoscale investigation of pathogenic microbial adhesion to a biomaterial. *Applied and Environmental Microbiology* 2004, 70 (10), 6012–22.

[152] Bowen, W. R.; Lovitt, R. W.; Wright, C. J., Atomic force microscopy study of the adhesion of Saccharomyces cerevisiae. *Journal of Colloid and Interface Science* 2001, 237 (1), 54–61.

[153] Janshoff, A.; Neitzert, M.; Oberdorfer, Y.; Fuchs, H., Force spectroscopy of molecular systems – Single molecule spectroscopy of polymers and biomolecules. *Angewandte Chemie International Edition* 2000, 39 (18), 3213–37.

[154] Butt, H.-J., Measuring local surface-charge densities in electrolyte-solutions with a scanning force microscope. *Biophysical Journal* 1992, 63 (2), 578–82.

[155] Dufrêne, Y. F.; Boonaert, C. J. P.; van der Mei, H. C.; Busscher, H. J.; Rouxhet, P. G., Probing molecular interactions and mechanical properties of microbial cell surfaces by atomic force microscopy. *Ultramicroscopy* 2001, 86 (1–2), 113–20.

[156] Dupres, V.; Menozzi, F. D.; Locht, C.; Clare, B. H.; Abbott, N. L.; Cuenot, S.; Bompard, C.; Raze, D.; Dufrêne, Y. F., Nanoscale mapping and functional analysis of individual adhesins on living bacteria. *Nature Methods* 2005, 2 (7), 515–20.

[157] Müller, D. J., AFM: a nanotool in membrane biology. *Biochemistry* 2008, 47 (31), 7986–98.

[158] Eaton, P.; Fernández Estarlich, F.; Ewen, R. J.; Nevell, T. G.; Smith, J. R.; Tsibouklis, J., Combined nanoindention and adhesion force mapping using the atomic force microscope: investigations of a filled polysiloxane coating. *Langmuir* 2002, 18 (25), 10011–15.

[159] Eaton, P.; Smith, J. R.; Graham, P.; Smart, J. D.; Nevell, T. G.; Tsibouklis, J., Adhesion force mapping of polymer surfaces: factors influencing force of adhesion. *Langmuir* 2002, 18 (8), 3387–89.

[160] Vancso, G. J.; Hillborg, H.; Schonherr, H., Chemical composition of polymer surfaces imaged by atomic force microscopy and complementary approaches. In *Polymer Analysis, Polymer Theory*, Springer, Berlin, 2005; Vol. 182, pp 55–129.

[161] Ducker, W. A.; Senden, T. J.; Pashley, R. M., Measurement of forces in liquids using a force microscope. *Langmuir* 1992, 8 (7), 1831–36.

[162] Gotzinger, M.; Peukert, W., Adhesion forces of spherical alumina particles on ceramic substrates. *Journal of Adhesion* 2004, 80 (3), 223–42.

[163] Lee, G. U.; Kidwell, D. A.; Colton, R. J., Sensing discrete streptavidin-biotin interactions with atomic force microscopy. *Langmuir* 1994, 10 (2), 354–7.

[164] Pelling, A. E.; Li, Y.; Shi, W.; Gimzewski, J. K., Nanoscale visualization and characterization of Myxococcus xanthus cells with atomic force microscopy. *Proceedings of the National Academy of Sciences of the United States of America* 2005, 102 (18), 6484–89.

[165] Duwez, A. S.; Poleunis, C.; Bertrand, P.; Nysten, B., Chemical recognition of antioxidants and uv-light stabilizers at the surface of polypropylene: atomic force microscopy with chemically modified tips. *Langmuir* 2001, 17 (20), 6351–57.

[166] Noy, A.; Vezenov, D. V.; Lieber, C. M., Chemical force microscopy. *Annual Review of Materials Science* 1997, 27, 381–421.

[167] VanLandingham, M. R., Review of instrumented indentation. *Journal of Research of the National Institute of Standards and Technology* 2003, 108 (4), 249–65.

[168] VanLandingham, M. R.; McKnight, S. H.; Palmese, G. R.; Elings, J. R.; Huang, X.; Bogetti, T. A.; Eduljee, R. F.; Gillespie, J. W., Nanoscale indentation of polymer systems using the atomic force microscope. *Journal of Adhesion* 1997, 64 (1–4), 31–59.

[169] Sirghi, L.; Rossi, F., Adhesion and elasticity in nanoscale indentation. *Applied Physics Letters* 2006, 89 (24), 243118–3.

[170] Eaton, P.; Fernandes, J. C.; Pereira, E.; Pintado, M. E.; Malcata, F. X., Atomic force microscopy study of the antibacterial effects of chitosans on *Escherichia coli* and *Staphylococcus aureus*. *Ultramicroscopy* 2008, 108 (10), 1128–34.

[171] Touhami, A.; Nysten, B.; Dufrêne, Y. F., Nanoscale mapping of the elasticity of microbial cells by atomic force microscopy. *Langmuir* 2003, 19 (11), 4539–43.

[172] Vadillo-Rodriguez, V.; Beveridge, T. J.; Dutcher, J. R., Surface viscoelasticity of individual Gram-negative bacterial cells measured using atomic force microscopy. *Journal of Bacteriology* 2008, 190 (12), 4225–32.

[173] Butt, H.-J., Measuring electrostatic, van der Waals, and hydration forces in electrolyte solutions with an atomic force microscope. *Biophysical Journal* 1991, 60 (6), 1438–44.

[174] Almqvist, N.; Delamo, Y.; Smith, B. L.; Thomson, N. H.; Bartholdson, Å.; Lal, R.; Brzezinski, M.; Hansma, P. K., Micromechanical and structural properties of a pennate diatom investigated by atomic force microscopy. *Journal of Microscopy* 2001, 202 (3), 518–32.

[175] Ludwig, T.; Kirmse, R.; Poole, K.; Schwarz, U., Probing cellular microenvironments and tissue remodeling by atomic force microscopy. *Pflügers Archiv European Journal of Physiology* 2008, 456 (1), 29–49.

[176] Carl, P.; Schillers, H., Elasticity measurement of living cells with an atomic force microscope: data acquisition and processing. *Pflügers Archiv European Journal of Physiology* 2008, 457 (2), 551–59.

[177] VanLandingham, M. R.; McKnight, S. H.; Palmese, G. R.; Eduljee, R. F.; Gillespie, J. W.; McCulough, R. L., Relating elastic modulus to indentation response using atomic force microscopy. *Journal of Materials Science Letters* 1997, 16 (2), 117–19.

[178] Fernandes, J. C.; Eaton, P.; Gomes, A. M.; Pintado, M. E.; Malcata, F. X., Study of the antibacterial effects of chitosans on *Bacillus cereus* (and its spores) by atomic force microscopy imaging and nanoindentation. *Ultramicroscopy* 2009, 109, 854–60.

[179] Gaillard, Y.; Tromas, C.; Woirgard, J., Quantitative analysis of dislocation pile-ups nucleated during nanoindentation in MgO. *Acta Materialia* 2006, 54 (5), 1409–17.

[180] Gaillard, Y.; Tromas, C.; Woirgard, J., Study of the dislocation structure involved in a nanoindentation test by atomic force microscopy and controlled chemical etching. *Acta Materialia* 2003, 51 (4), 1059–65.

[181] Bischel, M. S.; VanLandingham, M. R.; Eduljee, R. F.; Gillespie, J. W.; Schultz, J. M., On the use of nanoscale indentation with the AFM in the identification of phases in blends of linear low density polyethylene and high density polyethylene. *Journal of Materials Science* 2000, 35 (1), 221–28.

[182] VanLandingham, M. R.; Dagastine, R. R.; Eduljee, R. F.; McCullough, R. L.; Gillespie, J. W., Characterization of nanoscale property variations in polymer composite systems: 1. Experimental results. *Composites Part A – Applied Science and Manufacturing* 1999, 30 (1), 75–83.

[183] Penegar, I.; Toque, C.; Connell, S. D. A.; Smith, J. R.; Campbell, S. A., Nano-indentation measurements of the marine bacteria sphigomonas paucimobilis using the atomic force microscope. *Additional Papers from the 10th International Congress on Marine Corrosion and Fouling*, DSTO Aeronautical and Maritime Research Laboratory, Victoria, Australia, 2001.

[184] Volle, C. B.; Ferguson, M. A.; Aidala, K. E.; Spain, E. M.; Núñez, M. E., Spring constants and adhesive properties of native bacterial biofilm cells measured by atomic force microscopy. *Colloids and Surfaces B: Biointerfaces* 2008, 67 (1), 32–40.

[185] Wampler, H. P.; Ivanisevic, A., Nanoindentation of gold nanoparticles functionalized with proteins. *Micron* 2009, 40 (4), 444–48.

[186] Armini, S.; Vakarelski, I. U.; Whelan, C. M.; Maex, K.; Higashitani, K., Nanoscale indentation of polymer and composite polymer-silica core-shell submicrometer particles by atomic force microscopy. *Langmuir* 2007, 23 (4), 2007–14.

[187] Vakarelski, I. U.; Toritani, A.; Nakayama, M.; Higashitani, K., Effects of particle deformability on interaction between surfaces in solutions. *Langmuir* 2003, 19 (1), 110–17.

[188] Jeon, S.; Braiman, Y.; Thundat, T., Cross talk between bending, twisting, and buckling modes of three types of microcantilever sensors. *Review of Scientific Instruments* 2004, 75 (11), 4841–44.

[189] Gnecco, E.; Bennewitz, R.; Gyalog, T.; Meyer, E., Friction experiments on the nanometre scale. *Journal of Physics – Condensed Matter* 2001, 13 (31), R619–R642.

[190] Babcock, K. L.; Prater, C. B. *Phase Imaging: Beyond Topography*; Veeco Application Note: 1995.

[191] Tamayo, J.; García, R., Deformation, contact time, and phase contrast in tapping mode scanning force microscopy. *Langmuir* 1996, 12 (18), 4430–35.

[192] Schmitz, I.; Schreiner, M.; Friedbacher, G.; Grasserbauer, M., Phase imaging as an extension to tapping mode AFM for the identification of material properties on humidity-sensitive surfaces. *Applied Surface Science* 1997, 115 (2), 190–98.

[193] Nagao, E.; Dvorak, J. A., Phase imaging by atomic force microscopy: analysis of living homoiothermic vertebrate cells. *Biophysical Journal* 1999, 76 (6), 3289–97.

[194] Martínez, N. F.; García, R., Measuring phase shifts and energy dissipation with amplitude modulation atomic force microscopy. *Nanotechnology* 2006, 17 (7), S167–S172.

[195] Tamayo, J.; García, R., Relationship between phase shift and energy dissipation in tapping-mode scanning force microscopy. *Applied Physics Letters* 1998, 73 (20), 2926–28.

[196] García, R.; Tamayo, J.; San Paulo, A., Phase contrast and surface energy hysteresis in tapping mode scanning force microscopy. *Surface and Interface Analysis* 1999, 27 (5–6), 312–16.

[197] García, R.; Gomez, C. J.; Martinez, N. F.; Patil, S.; Dietz, C.; Magerle, R., Identification of nanoscale dissipation processes by dynamic atomic force microscopy. *Physical Review Letters* 2006, 97 (1), 016103.

[198] de Pablo, P. J.; Colchero, J.; Gomez-Herrero, J.; Baro, A. M., Jumping mode scanning force microscopy. *Applied Physics Letters* 1998, 73 (22), 3300–2.

[199] Rosa-Zeiser, A.; Weilandt, E.; Hild, S.; Marti, O., The simultaneous measurement of elastic, electrostatic and adhesive properties by scanning force microscopy: pulsed-force mode operation. *Measurement Science and Technology* 1997, 8 (11), 1333–38.

[200] Sotres, J.; Lostao, A.; Gómez-Moreno, C.; Baró, A. M., Jumping mode AFM imaging of biomolecules in the repulsive electrical double layer. *Ultramicroscopy* 2007, 107, 1207–12.

[201] Moreno-Herrero, F.; de Pablo, P. J.; Colchero, J.; Gómez-Herrero, J.; Baró, A. M., The role of shear forces in scanning force microscopy: a comparison between the jumping mode and tapping mode. *Surface Science* 2000, 453 (1–3), 152–58.

[202] Moreno-Herrero, F.; Colchero, J.; Gómez-Herrero, J.; Baró, A. M., Atomic force microscopy contact, tapping, and jumping modes for imaging biological samples in liquids. *Physical Review E* 2004, 69 (3), 031915.

[203] Moreno-Herrero, F.; de Pablo, P. J.; Alvarez, M.; Colchero, J.; Gómez-Hertero, J.; Baró, A. M., Jumping mode scanning force microscopy: a suitable technique for imaging DNA in liquids. *Applied Surface Science* 2003, 210 (1–2), 22–26.

[204] Moreno-Herrero, F.; de Pablo, P. J.; Fernandez-Sanchez, R.; Colchero, J.; Gómez-Herrero, J.; Baró, A. M., Scanning force microscopy jumping and tapping modes in liquids. *Applied Physics Letters* 2002, 81 (14), 2620–22.

[205] Krüger, S.; Krüger, D.; Janshoff, A., Scanning force microscopy based rapid force curve acquisition on supported lipid bilayers: experiments and simulations using pulsed force mode. *ChemPhysChem* 2004, 5 (7), 989–97.

[206] Krotil, H.-U.; Stifter, T.; Waschipky, H.; Weishaupt, K.; Hild, S.; Marti, O., Pulsed force mode: a new method for the investigation of surface properties. *Surface and Interface Analysis* 1999, 27 (5–6), 336–40.

[207] Holzwarth, M. J.; Gigler, A. M.; Marti, O., Digital pulsed force mode. *Imaging & Microscopy* 2006, 8 (4), 37–38.

[208] Vanderwerf, K. O.; Putman, C. A. J.; Degrooth, B. G.; Greve, J., Adhesion force imaging in air and liquid by adhesion mode atomic-force microscopy. *Applied Physics Letters* 1994, 65 (9), 1195–97.

[209] Martin, Y.; Rugar, D.; Wickramasinghe, H. K., High-resolution magnetic imaging of domains in TbFe by force microscopy. *Applied Physics Letters* 1988, 52 (3), 244–46.

[210] Mamin, H. J.; Rugar, D.; Stern, J. E.; Terris, B. D.; Lambert, S. E., Force microscopy of magnetization patterns in longitudinal recording media. *Applied Physics Letters* 1988, 53 (16), 1563–65.

[211] Hartmann, U., Magnetic force microscopy. *Annual Review of Materials Science* 1999, 29, 53–87.

[212] Abelmann, L.; Porthun, S.; Haast, M.; Lodder, C.; Moser, A.; Best, M. E.; van Schendel, P. J. A.; Stiefel, B.; Hug, H. J.; Heydon, G. P.; Farley, A.; Hoon, S. R.; Pfaffelhuber, T.; Proksch, R.; Babcock, K., Comparing the resolution of magnetic force microscopes using the CAMST reference samples. *Journal of Magnetism and Magnetic Materials* 1998, 190 (1–2), 135–47.

[213] Porthun, S.; Abelmann, L.; Lodder, C., Magnetic force microscopy of thin film media for high density magnetic recording. *Journal of Magnetism and Magnetic Materials* 1998, 182 (1–2), 238–73.

[214] Lin, C. W.; Fan, F.-R. F.; Bard, A. J., High resolution photoelectrochemical etching of n-GaAs with the scanning electrochemical and tunneling microscope. *Journal of the Electrochemical Society* 1987, 134 (4), 1038–39.

[215] Giles, R.; Cleveland, J. P.; Manne, S.; Hansma, P. K.; Drake, B.; Maivald, P.; Boles, C.; Gurley, J.; Elings, V., Noncontact force microscopy in liquids. *Applied Physics Letters* 1993, 63 (5), 617–18.

[216] Hosaka, S.; Kikukawa, A.; Honda, Y.; Hasegawa, T., Just-on-surface magnetic force microscopy. *Applied Physics Letters* 1994, 65 (26), 3407–9.

[217] Raşa, M.; Kuipers, B. W. M.; Philipse, A. P., Atomic force microscopy and magnetic force microscopy study of model colloids. *Journal of Colloid and Interface Science* 2002, 250 (2), 303–15.

[218] Proksch, R.; Skidmore, G. D.; Dahlberg, E. D.; Foss, S.; Schmidt, J. J.; Merton, C.; Walsh, B.; Dugas, M., Quantitative magnetic field measurements with the magnetic force microscope. *Applied Physics Letters* 1996, 69 (17), 2599–2601.

[219] Proksch, R., Recent advances in magnetic force microscopy. *Current Opinion in Solid State & Materials Science* 1999, 4 (2), 231–36.

[220] Suter, A., The magnetic resonance force microscope. *Progress in Nuclear Magnetic Resonance Spectroscopy* 2004, 45 (3–4), 239–74.

[221] Schreiber, S.; Savla, M.; Pelekhov, D. V.; Iscru, D. F.; Selcu, C.; Hammel, P. C.; Agarwal, G., Magnetic force microscopy of superparamagnetic nanoparticles. *Small* 2008, 4 (2), 270–78.

[222] Stern, J. E.; Terris, B. D.; Mamin, H. J.; Rugar, D., Deposition and imaging of localized charge on insulator surfaces using a force microscope. *Applied Physics Letters* 1988, 53 (26), 2717–19.

[223] Ratzke, M.; Reif, H., On the reliability of scanning probe based electrostatic force measurements. *Microelectronic Engineering* 2007, 84 (3), 512–16.

[224] Nonnenmacher, M.; Oboyle, M. P.; Wickramasinghe, H. K., Kelvin probe force microscopy. *Applied Physics Letters* 1991, 58 (25), 2921–23.

[225] Palermo, V.; Palma, M.; Samori, P., Electronic characterization of organic thin films by Kelvin probe force microscopy. *Advanced Materials* 2006, 18 (2), 145–64.

[226] Jacobs, H. O.; Leuchtmann, P.; Homan, O. J.; Stemmer, A., Resolution and contrast in Kelvin probe force microscopy. *Journal of Applied Physics* 1998, 84 (3), 1168–73.

[227] Cui, X.; Freitag, M.; Martel, R.; Brus, L.; Avouris, P., Controlling energy-level alignments at carbon nanotube/Au contacts. *Nano Letters* 2003, 3 (6), 783–87.

[228] Hillner, P. E.; Manne, S.; Gratz, A. J.; Hansma, P. K., AFM images of dissolution and growth on a calcite crystal. *Ultramicroscopy* 1992, 42, 1387–93.

[229] Macpherson, J. V.; Unwin, P. R., Combined scanning electrochemical-atomic force microscopy. *Analytical Chemistry* 2000, 72 (2), 276–85.

[230] Vidu, R.; Ku, J.-R.; Stroeve, P., Growth of ultrathin films of cadmium telluride and tellurium as studied by electrochemical atomic force microscopy. *Journal of Colloid and Interface Science* 2006, 300 (1), 404–12.

[231] Pollock, H. M.; Hammiche, A., Micro-thermal analysis: techniques and applications. *Journal of Physics D: Applied Physics* 2001, 34 (9), R23–R53.

[232] Price, D. M.; Reading, M.; Hammiche, A.; Pollock, H. M., Micro-thermal analysis: scanning thermal microscopy and localised thermal analysis. *International Journal of Pharmaceutics* 1999, 192 (1), 85–96.

[233] Hammiche, A.; Price, D. M.; Dupas, E.; Mills, G.; Kulik, A.; Reading, M.; Weaver, J. M. R.; Pollock, H. M., Two new microscopical variants of thermomechanical modulation: scanning thermal expansion microscopy and dynamic localized thermomechanical analysis. *Journal of Microscopy* 2000, 199 (3), 180–90.

[234] Tsukruk, V. V.; Gorbunov, V. V.; Fuchigami, N., Microthermal analysis of polymeric materials. *Thermochimica Acta* 2003, 395 (1–2), 151–58.

[235] Fischer, H., Calibration of micro-thermal analysis for the detection of glass transition temperatures and melting points –repeatability and reproducibility. *Journal of Thermal Analysis and Calorimetry* 2008, 92 (2), 625–30.

[236] Harding, L.; King, W. P.; Dai, X.; Craig, D. Q. M.; Reading, M., Nanoscale characterisation and imaging of partially amorphous materials using local thermomechanical analysis and heated tip AFM. *Pharmaceutical Research* 2007, 24 (11), 2048–54.

[237] Zhang, J. X.; Roberts, C. J.; Shakesheff, K. M.; Davies, M. C.; Tendler, S. J. B., Micro- and macrothermal analysis of a bioactive surface-engineered polymer formed by physical entrapment of poly(ethylene glycol) into poly(lactic acid). *Macromolecules* 2003, 36 (4), 1215–21.

[238] Mallarino, S.; Chailan, J. F.; Vernet, J. L., Interphase investigation in glass fibre composites by micro-thermal analysis. *Composites Part A –Applied Science and Manufacturing* 2005, 36 (9), 1300–6.

[239] Woodward, I.; Ebbens, S.; Zhang, J.; Luk, S.; Patel, N.; Roberts, C. J., A combined imaging, microthermal and spectroscopic study of a multilayer packaging system. *Packaging Technology and Science* 2004, 17 (3), 129–38.

[240] Bond, L.; Allen, S.; Davies, M. C.; Roberts, C. J.; Shivji, A. P.; Tendler, S. J. B.; Williams, P. M.; Zhang, J. X., Differential scanning calorimetry and scanning thermal microscopy analysis of pharmaceutical materials. *International Journal of Pharmaceutics* 2002, 243 (1–2), 71–82.

[241] Germinario, L. T.; Shang, P. P., Advances in nano thermal analysis of coatings. *Journal of Thermal Analysis and Calorimetry* 2008, 93 (1), 207–11.

[242] Zhang, J. X.; Botterill, N. W.; Roberts, C. J.; Grant, D. M., Micro-thermal analysis of NiTi shape memory alloy thin films. *Thermochimica Acta* 2003, 401 (2), 111–19.

[243] Boroumand, F. A.; Hammiche, A.; Hill, G.; Lidzey, D. G., Characterizing joule heating in polymer light-emitting diodes using a scanning thermal microscope. *Advanced Materials* 2004, 16 (3), 252–56.

[244] Stievenard, D.; Legrand, B., Silicon surface nano-oxidation using scanning probe microscopy. *Progress in Surface Science* 2006, 81 (2–3), 112–40.

[245] Rank, R.; Bruckl, H.; Kretz, J.; Monch, I.; Reiss, G., Nanoscale modification of conducting lines with a scanning force microscope. *Vacuum* 1997, 48 (5), 467–72.

[246] Ginger, D. S.; Zhang, H.; Mirkin, C. A., The evolution of dip-pen nanolithography. *Angewandte Chemie International Edition* 2004, 43 (1), 30–45.

[247] Quate, C. F., Manipulation and modification of nanometer scale objects with the STM. *Highlights in Condensed Matter Physics and Future Prospects* 1991, 285, 573–630.

[248] Day, H. C.; Allee, D. R., Selective area oxidation of silicon with a scanning force microscope. *Applied Physics Letters* 1993, 62 (21), 2691–93.

[249] Xie, X. N.; Chung, H. J.; Sow, C. H.; Wee, A. T. S., Nanoscale materials patterning and engineering by atomic force microscopy nanolithography. *Materials Science & Engineering R-Reports* 2006, 54 (1–2), 1–48.

[250] Martínez, R. V.; Losilla, N. S.; Martinez, J.; Huttel, Y.; García, R., Patterning polymeric structures with 2 nm resolution at 3 nm half pitch in ambient conditions. *Nano Letters* 2007, 7 (7), 1846–50.

[251] Snow, E. S.; Campbell, P. M.; Perkins, F. K., Nanofabrication with proximal probes. *Proceedings of the IEEE* 1997, 85 (4), 601–11.

[252] Snow, E. S.; Campbell, P. M.; Perkins, F. K., High speed patterning of a metal silicide using scanned probe lithography. *Applied Physics Letters* 1999, 75 (10), 1476–78.

[253] Fontaine, P. A.; Dubois, E.; Stievenard, D., Characterization of scanning tunneling microscopy and atomic force microscopy-based techniques for nanolithography on hydrogen-passivated silicon. *Journal of Applied Physics* 1998, 84 (4), 1776–81.

[254] Tello, M.; García, R., Nano-oxidation of silicon surfaces: comparison of noncontact and contact atomic-force microscopy methods. *Applied Physics Letters* 2001, 79 (3), 424–26.

[255] Cavallini, M.; Mei, P.; Biscarini, F.; García, R., Parallel writing by local oxidation nanolithography with submicrometer resolution. *Applied Physics Letters* 2003, 83 (25), 5286–88.

[256] Martínez, R. V.; Losilla, N. S.; Martinez, J.; Tello, M.; García, R., Sequential and parallel patterning by local chemical nanolithography. *Nanotechnology* 2007, 18 (8), 084021.

[257] Minne, S. C.; Adams, J. D.; Yaralioglu, G.; Manalis, S. R.; Atalar, A.; Quate, C. F., Centimeter scale atomic force microscope imaging and lithography. *Applied Physics Letters* 1998, 73 (12), 1742–44.

[258] Held, R.; Vancura, T.; Heinzel, T.; Ensslin, K.; Holland, M.; Wegscheider, W., In-plane gates and nanostructures fabricated by direct oxidation of semiconductor heterostructures with an atomic force microscope. *Applied Physics Letters* 1998, 73 (2), 262–64.

[259] Held, R.; Heinzel, T.; Studerus, P.; Ensslin, K.; Holland, M., Semiconductor quantum point contact fabricated by lithography with an atomic force microscope. *Applied Physics Letters* 1997, 71 (18), 2689–91.

[260] Piner, R. D.; Zhu, J.; Xu, F.; Hong, S. H.; Mirkin, C. A., 'Dip-pen' nanolithography. *Science* 1999, 283 (5402), 661–63.

[261] Weeks, B. L.; Noy, A.; Miller, A. E.; De Yoreo, J. J., Effect of dissolution kinetics on feature size in dip-pen nanolithography. *Physical Review Letters* 2002, 88 (25), 255505.

[262] Christman, K. L.; Enriquez-Rios, V. D.; Maynard, H. D., Nanopatterning proteins and peptides. *Soft Matter* 2006, 2 (11), 928–39.

[263] Noy, A.; Miller, A. E.; Klare, J. E.; Weeks, B. L.; Woods, B. W.; DeYoreo, J. J., Fabrication of luminescent nanostructures and polymer nanowires using dip-pen nanolithography. *Nano Letters* 2002, 2 (2), 109–12.

[264] Agarwal, G.; Sowards, L. A.; Naik, R. R.; Stone, M. O., Dip-pen nanolithography in tapping mode. *Journal of the American Chemical Society* 2003, 125 (2), 580–83.

[265] Fu, L.; Liu, X.; Zhang, Y.; Dravid, V. P.; Mirkin, C. A., Nanopatterning of 'hard' magnetic nanostructures via dip-pen nanolithography and a sol-based ink. *Nano Letters* 2003, 3 (6), 757–60.

[266] Tseng, A. A.; Notargiacomo, A.; Chen, T. P., Nanofabrication by scanning probe microscope lithography: a review. *Journal of Vacuum Science & Technology B* 2005, 23 (3), 877–94.

[267] King, W. P.; Kenny, T. W.; Goodson, K. E.; Cross, G.; Despont, M.; Durig, U.; Rothuizen, H.; Binnig, G. K.; Vettiger, P., Atomic force microscope cantilevers for combined thermomechanical data writing and reading. *Applied Physics Letters* 2001, 78 (9), 1300–2.

[268] Durig, U.; Cross, G.; Despont, M.; Drechsler, U.; Haberle, W.; Lutwyche, M. I.; Rothuizen, H.; Stutz, R.; Widmer, R.; Vettiger, P.; Binnig, G. K.; King, W. P.; Goodson, K. E., 'Millipede' – an AFM data storage system at the frontier of nanotribology. *Tribology Letters* 2000, 9 (1–2), 25–32.

[269] Pozidis, H.; Haberle, W.; Wiesmann, D.; Drechsler, U.; Despont, M.; Albrecht, T. R.; Eleftheriou, E., Demonstration of thermomechanical recording at 641 Gbit/in^2. *IEEE Transactions on Magnetics* 2004, 40 (4), 2531–36.

[270] Junno, T.; Deppert, K.; Montelius, L.; Samuelson, L., Controlled manipulation of nanoparticles with an atomic-force microscope. *Applied Physics Letters* 1995, 66 (26), 3627–29.

[271] Hu, J.; Zhang, Y.; Gao, H. B.; Li, M. Q.; Hartmann, U., Artificial DNA patterns by mechanical nanomanipulation. *Nano Letters* 2002, 2 (1), 55–57.

[272] Ternes, M.; Lutz, C. P.; Hirjibehedin, C. F.; Giessibl, F. J.; Heinrich, A. J., The force needed to move an atom on a surface. *Science* 2008, 319 (5866), 1066–69.

[273] Sugimoto, Y.; Abe, M.; Hirayama, S.; Oyabu, N.; Custance, O.; Morita, S., Atom inlays performed at room temperature using atomic force microscopy. *Nature Materials* 2005, 4 (2), 156–59.

[274] Liu, M.; Amro, N. A.; Liu, G. Y., Nanografting for surface physical chemistry. *Annual Review of Physical Chemistry* 2008, 59, 367–86.

[275] Liang, J.; Scoles, G., Nanografting of alkanethiols by tapping mode atomic force microscopy. *Langmuir* 2007, 23 (11), 6142–47.

[276] Wadu-Mesthrige, K.; Xu, S.; Amro, N. A.; Liu, G. Y., Fabrication and imaging of nanometer-sized protein patterns. *Langmuir* 1999, 15 (25), 8580–83.

[277] Xu, S.; Liu, G. Y., Nanometer-scale fabrication by simultaneous nanoshaving and molecular self-assembly. *Langmuir* 1997, 13 (2), 127–29.

[278] Paik, P.; Kar, K. K.; Deva, D.; Sharma, A., Measurement of mechanical properties of polymer nanospheres by atomic force microscopy: effects of particle size. *Micro & Nano Letters* 2007, 2 (3), 72–77.

[279] Abdelhady, H. G.; Allen, S.; Ebbens, S. J.; Madden, C.; Patel, N.; Roberts, C. J.; Zhang, J. X., Towards nanoscale metrology for biomolecular imaging by atomic force microscopy. *Nanotechnology* 2005, 16 (6), 966–73.

[280] De la Fuente, J. M.; Eaton, P.; Barrientos, A. G.; Menendez, M.; Penades, S., Thermodynamic evidence for Ca^{2+}-mediated self-aggregation of Lewis X gold glyconanoparticles. A model for cell adhesion via carbohydrate–carbohydrate interaction. *Journal of the American Chemical Society* 2005, 127 (17), 6192–97.

[281] Wang, H.; Tessmer, I.; Croteau, D. L.; Erie, D. A.; Van Houten, B., Functional characterization and atomic force microscopy of a DNA repair protein conjugated to a quantum dot. *Nano Letters* 2008, 8 (6), 1631–37.

[282] Quintana, M.; Haro-Poniatowski, E.; Morales, J.; Batina, N., Synthesis of selenium nanoparticles by pulsed laser ablation. *Applied Surface Science* 2002, 195 (1–4), 175–86.

[283] Pimpha, N.; Rattanonchai, U.; Surassmo, S.; Opanasopit, P.; Rattanarungchai, C.; Sunintaboon, P., Preparation of PMMA/acid-modified chitosan core-shell nanoparticles and their potential as gene carriers. *Colloid and Polymer Science* 2008, 286 (8–9), 907–16.

[284] Garcia, P.; Eaton, P.; Geurts, H. P. M.; Sousa, M.; Gameiro, P.; Feiters, M. C.; Nolte, R. J. M.; Pereira, E.; de Castro, B., AFM and electron microscopy study of the unusual aggregation behavior of metallosurfactants based on iron(II) complexes with bipyridine ligands. *Langmuir* 2007, 23 (15), 7951–57.

[285] Thomson, N. H., Imaging the substructure of antibodies with tapping-mode AFM in air: the importance of a water layer on mica. *Journal of Microscopy* 2005, 217 (3), 193–99.

[286] Bickmore, B. R.; Hochella, M. F.; Bosbach, D.; Charlet, L., Methods for performing atomic force microscopy imaging of clay minerals in aqueous solutions. *Clays and Clay Minerals* 1999, 47 (5), 573–81.

[287] Vesenka, J.; Manne, S.; Giberson, R.; Marsh, T.; Henderson, E., Colloidal gold particles as an incompressible atomic-force microscope imaging standard for assessing the compressibility of biomolecules. *Biophysical Journal* 1993, 65 (3), 992–97.

[288] Doron, A.; Joselevich, E.; Schlittner, A.; Willner, I., AFM characterization of the structure of Au-colloid monolayers and their chemical etching. *Thin Solid Films* 1999, 340 (1–2), 183–88.

[289] Sato, H.; Ohtsu, T.; Komasawa, I., Atomic force microscopy study of ultrafine particles prepared in reverse micelles. *Journal of Colloid and Interface Science* 2000, 230 (1), 200–4.

[290] Vinelli, A.; Primiceri, E.; Brucale, M.; Zuccheri, G.; Rinaldi, R.; Samori, B., Sample preparation for the quick sizing of metal nanoparticles by atomic force microscopy. *Microscopy Research and Technique* 2008, 71 (12), 870–79.

[291] Benoit, M.; Holstein, T.; Gaub, H. E., Lateral forces in AFM imaging and immobilization of cells and organelles. *European Biophysics Journal* 1997, 26 (4), 283–90.

[292] Lobo, R. F. M.; Pereira-da-Silva, M. A.; Raposo, M.; Faria, R. M.; Oliveira Jr., O. N., *In situ* thickness measurements of ultra-thin multilayer polymer films by atomic force microscopy. *Nanotechnology* 1999, 10 (4), 389–93.

[293] Kaczmarek, H.; Chaberska, H., The influence of solvent residue, support type and UV-irradiation on surface morphology of poly(methyl methacrylate) films studied by atomic force microscopy. *Polymer Testing* 2008, 27 (6), 736–42.

[294] Wagner, P., Immobilization strategies for biological scanning probe microscopy. *FEBS Letters* 1998, 430 (1–2), 112–15.

[295] Adamcik, J.; Klinov, D. V.; Witz, G.; Sekatskii, S. K.; Dietler, G., Observation of single-stranded DNA on mica and highly oriented pyrolytic graphite by atomic force microscopy. *FEBS Letters* 2006, 580 (24), 5671–75.

[296] Brett, A. M. O.; Chiorcea, A. M., Atomic force microscopy of DNA immobilized onto a highly oriented pyrolytic graphite electrode surface. *Langmuir* 2003, 19 (9), 3830–39.

[297] Vesenka, J.; Guthold, M.; Tang, C. L.; Keller, D.; Delaine, E.; Bustamante, C., Substrate preparation for reliable imaging of DNA-molecules with the scanning force microscope. *Ultramicroscopy* 1992, 42, 1243–49.

[298] Hansma, H. G.; Laney, D. E., DNA binding to mica correlates with cationic radius: assay by atomic force microscopy. *Biophysical Journal* 1996, 70 (4), 1933–39.

[299] Sanchez-Sevilla, A.; Thimonier, J.; Marilley, M.; Rocca-Serra, J.; Barbet, J., Accuracy of AFM measurements of the contour length of DNA fragments adsorbed on mica in air and in aqueous buffer. *Ultramicroscopy* 2002, 92 (3–4), 151–58.

[300] Lyubchenko, Y. L.; Shlyakhtenko, L. S., AFM for analysis of structure and dynamics of DNA and protein-DNA complexes. *Methods* 2009, 47 (3), 206–13.

[301] Gaczynska, M.; Osmulski, P. A., AFM of biological complexes: what can we learn? *Current Opinion in Colloid & Interface Science* 2008, 13 (5), 351–67.

[302] El Kirat, K.; Burton, I.; Dupres, V.; Dufrêne, Y. F., Sample preparation procedures for biological atomic force microscopy. *Journal of Microscopy* 2005, 218 (3), 199–207.

[303] Cullen, D. C.; Lowe, C. R., AFM studies of protein adsorption. 1. Time-resolved protein adsorption to highly oriented pyrolytic-graphite. *Journal of Colloid and Interface Science* 1994, 166 (1), 102–8.

[304] Viani, M. B.; Pietrasanta, L. I.; Thompson, J. B.; Chand, A.; Gebeshuber, I. C.; Kindt, J. H.; Richter, M.; Hansma, H. G.; Hansma, P. K., Probing protein–protein interactions in real time. *Nature Structural Biology* 2000, 7 (8), 644–47.

[305] Czajkowsky, D. M.; Shao, Z., Inhibition of protein adsorption to muscovite mica by monovalent cations. *Journal of Microscopy* 2003, 211 (1), 1–7.

[306] Klein, D. C. G.; Stroh, C. M.; Jensenius, H.; van Es, M.; Kamruzzahan, A. S. M.; Stamouli, A.; Gruber, H. J.; Oosterkamp, T. H.; Hinterdorfer, P., Covalent immobilization of single proteins on mica for molecular recognition force microscopy. *ChemPhysChem* 2003, 4 (12), 1367–71.

[307] Müller, D. J.; Engel, A., Strategies to prepare and characterize native membrane proteins and protein membranes by AFM. *Current Opinion in Colloid & Interface Science* 2008, 13 (5), 338–50.

[308] Neves, P.; Lopes, S. C. D. N.; Sousa, I.; Garcia, S.; Eaton, P.; Gameiro, P., Characterization of membrane protein reconstitution in LUVs of different lipid composition by fluorescence anisotropy. *Journal of Pharmaceutical and Biomedical Analysis* 2008, 49 (2), 276–81.

[309] Pelling, A. E.; Veraitch, F. S.; Chu, C. P.-K.; Nicholls, B. M.; Hemsley, A. L.; Mason, C.; Horton, M. A., Mapping correlated membrane pulsations and fluctuations in human cells. *Journal of Molecular Recognition* 2007, 20 (6), 467–75.

[310] Espenel, C.; Giocondi, M.-C.; Seantier, B.; Dosset, P.; Milhiet, P.-E.; Le Grimellec, C., Temperature-dependent imaging of living cells by AFM. *Ultramicroscopy* 2008, 108 (10), 1174–80.

[311] Braet, F.; de Zanger, R.; Seynaeve, C.; Baekeland, M.; Wisse, E., A comparative atomic force microscopy study on living skin fibroblasts and liver endothelial cells. *Journal of Electron Microscopy* 2001, 50 (4), 283–90.

[312] Núñez, M. E.; Martin, M. O.; Chan, P. H.; Duong, L. K.; Sindhurakar, A. R.; Spain, E. M., Atomic force microscopy of bacterial communities. In *Environmental Microbiology*, Elsevier Academic Press Inc: San Diego, 2005; Vol. 397, pp 256–68.

[313] Vadillo-Rodriguez, V.; Busscher, H. J.; Norde, W.; de Vries, J.; Dijkstra, R. J. B.; Stokroos, I.; van der Mei, H. C., Comparison of atomic force microscopy interaction forces between bacteria and silicon nitride substrata for three commonly used immobilization methods. *Applied and Environmental Microbiology* 2004, 70 (9), 5541–46.

[314] Burks, G. A.; Velegol, S. B.; Paramonova, E.; Lindenmuth, B. E.; Feick, J. D.; Logan, B. E., Macroscopic and nanoscale measurements of the adhesion of bacteria with varying outer layer surface composition. *Langmuir* 2003, 19 (6), 2366–71.

[315] Micic, M.; Hu, D.; Suh, Y. D.; Newton, G.; Romine, M.; Lu, H. P., Correlated atomic force microscopy and fluorescence lifetime imaging of live bacterial cells. *Colloids and Surfaces B: Biointerfaces* 2004, 34 (4), 205–12.

[316] Volle, C. B.; Ferguson, M. A.; Aidala, K. E.; Spain, E. M.; Nuñez, M. E., Quantitative changes in the elasticity and adhesive properties of *Escherichia coli* ZK1056 prey cells during predation by *Bdellovibrio bacteriovorus* 109J. *Langmuir* 2008, 24 (15), 8102–10.

[317] Kailas, L.; Ratcliffe, E. C.; Hayhurst, E. J.; Walker, M. G.; Foster, S. J.; Hobbs, J. K., Immobilizing live bacteria for AFM imaging of cellular processes. *Ultramicroscopy* 2009, 109 (7), 775–80.

[318] Schultze, J. W.; Davepon, B.; Karman, F.; Rosenkranz, C.; Schreiber, A.; Voigt, O., Corrosion and passivation in nanoscopic and microscopic dimensions: the influence of grains and grain boundaries. *Corrosion Engineering, Science and Technology* 2004, 39, 45–52.

[319] Bezanilla, M.; Manne, S.; Laney, D. E.; Lyubchenko, Y. L.; Hansma, H. G., Adsorption of DNA to mica, silylated mica, and minerals – characterization by atomic-force microscopy. *Langmuir* 1995, 11 (2), 655–59.

[320] Müller, D. J.; Janovjak, H.; Lehto, T.; Kuerschner, L.; Anderson, K., Observing structure, function and assembly of single proteins by AFM. *Progress in Biophysics and Molecular Biology* 2002, 79 (1–3), 1–43.

[321] Bystrenova, E.; Radenovic, A.; Libioulle, L.; Dietler, G. *Importance of substrate for bio-logical imaging by AFM at low temperature. Scanning tunneling microscopy/spectroscopy and related techniques: 12th International Conference.* AIP: Eindhoven, Netherlands, 2003; pp 461–66.

[322] Bhushan, B.; Tokachichu, D. R.; Keener, M. T.; Lee, S. C., Morphology and adhesion of biomolecules on silicon based surfaces. *Acta Biomaterialia* 2005, 1 (3), 327–41.

[323] Olbrich, A.; Ebersberger, B.; Boit, C., Conducting atomic force microscopy for nanoscale electrical characterization of thin SiO2. *Applied Physics Letters* 1998, 73 (21), 3114–16.

[324] Henke, L.; Nagy, N.; Krull, U. J., An AFM determination of the effects on surface roughness caused by cleaning of fused silica and glass substrates in the process of optical biosensor preparation. *Biosensors & Bioelectronics* 2002, 17 (6–7), 547–55.

[325] Banner, L. T.; Richter, A.; Pinkhassik, E., Pinhole-free large-grained atomically smooth Au (111) substrates prepared by flame-annealed template stripping. *Surface and Interface Analysis* 2009, 41 (1), 49–55.

[326] Rundqvist, J.; Hoh, J. H.; Haviland, D. B., Substrate effects in poly(ethylene glycol) self-assembled monolayers on granular and flame-annealed gold. *Journal of Colloid and Interface Science* 2006, 301 (1), 337–41.

[327] Rundqvist, J.; Hoh, J. H.; Haviland, D. B., Poly(ethylene glycol) self-assembled monolayer island growth. *Langmuir* 2005, 21 (7), 2981–87.

[328] Hegner, M.; Wagner, P.; Semenza, G., Ultralarge atomically flat template-stripped Au surfaces for scanning probe microscopy. *Surface Science* 1993, 291 (1–2), 39–46.

[329] Wagner, P.; Hegner, M.; Guntherodt, H. J.; Semenza, G., Formation and *in-situ* modification of monolayers chemisorbed on ultraflat template-stripped gold surfaces. *Langmuir* 1995, 11 (10), 3867–75.

[330] Mosley, D. W.; Chow, B. Y.; Jacobson, J. M., Solid-state bonding technique for template-stripped ultraflat gold Substrates. *Langmuir* 2006, 22 (6), 2437–40.

[331] Cannara, R. J.; Eglin, M.; Carpick, R. W., Lateral force calibration in atomic force microscopy: a new lateral force calibration method and general guidelines for optimization. *Review of Scientific Instruments* 2006, 77 (5), 053701.

[332] Cain, R. G.; Reitsma, M. G.; Biggs, S.; Page, N. W., Quantitative comparison of three calibration techniques for the lateral force microscope. *Review of Scientific Instruments* 2001, 72 (8), 3304–12.

[333] Prunici, P.; Hess, P., Quantitative characterization of crosstalk effects for friction force microscopy with scan-by-probe SPMs. *Ultramicroscopy* 2008, 108 (7), 642–45.

[334] Leggett, G. J.; Brewer, N. J.; Chonga, K. S. L., Friction force microscopy: towards quantitative analysis of molecular organisation with nanometre spatial resolution. *Physical Chemistry Chemical Physics* 2005, 7 (6), 1107–20.

[335] Xie, H.; Vitard, J.; Haliyo, S.; Regnier, S.; Boukallel, M., Calibration of lateral force measurements in atomic force microscopy with a piezoresistive force sensor. *Review of Scientific Instruments* 2008, 79 (3), 033708.

[336] Cain, R. G.; Biggs, S.; Page, N. W., Force calibration in lateral force microscopy. *Journal of Colloid and Interface Science* 2000, 227 (1), 55–65.

[337] Varenberg, M.; Etsion, I.; Halperin, G., An improved wedge calibration method for lateral force in atomic force microscopy. *Review of Scientific Instruments* 2003, 74 (7), 3362–67.

[338] Ecke, S.; Raiteri, R.; Bonaccurso, E.; Reiner, C.; Deiseroth, H.-J.; Butt, H. J., Measuring normal and friction forces acting on individual fine particles. *Review of Scientific Instruments* 2001, 72 (11), 4164–70.

[339] Asay, D. B.; Kim, S. H., Direct force balance method for atomic force microscopy lateral force calibration. *Review of Scientific Instruments* 2006, 77 (4), 043903.

[340] Schäffer, T. E.; Radmacher, M.; Proksch, R., Magnetic force gradient mapping. *Journal of Applied Physics* 2003, 94 (10), 6525–32.

[341] García, R.; San Paulo, A., Dynamics of a vibrating tip near or in intermittent contact with a surface. *Physical Review B* 2000, 61 (20), 13381–84.

[342] García, R.; San Paulo, A., Amplitude curves and operating regimes in dynamic atomic force microscopy. *Ultramicroscopy* 2000, 82 (1–4), 79–83.

[343] Motamedi, R.; Wood-Adams, P. M., Influence of fluid cell design on the frequency response of AFM microcantilevers in liquid media. *Sensors* 2008, 8 (9), 5927–41.

[344] Kokavecz, J.; Mechler, A., Investigation of fluid cell resonances in intermittent contact mode atomic force microscopy. *Applied Physics Letters* 2007, 91 (2), 023113.

[345] Nnebe, I.; Schneider, J. W., Characterization of distance-dependent damping in tapping-mode atomic force microscopy force measurements in liquid. *Langmuir* 2004, 20 (8), 3195–3201.

[346] Tamayo, J.; Humphris, A. D. L.; Miles, M. J., Piconewton regime dynamic force microscopy in liquid. *Applied Physics Letters* 2000, 77 (4), 582–84.

[347] Kowalewski, T.; Holtzman, D. M. *In situ* atomic force microscopy study of Alzheimer's beta-amyloid peptide on different substrates: new insights into mechanism of beta-sheet formation, *Proceedings of the National Academy of Sciences, of the United States of America* 1999, 96 (7), 3688–93.

[348] Herruzo, E. T.; García, R., Frequency response of an atomic force microscope in liquids and air: magnetic versus acoustic excitation. *Applied Physics Letters* 2007, 91 (14), 143113.

[349] Putman, C. A. J.; Vanderwerf, K. O.; Degrooth, B. G.; Vanhulst, N. F.; Greve, J., Viscoelasticity of living cells allows high resolution imaging by tapping mode atomic force microscopy. *Biophysical Journal* 1994, 67 (4), 1749–53.

[350] Maali, A.; Hurth, C.; Cohen-Bouhacina, T.; Couturier, G.; Aime, J. P., Improved acoustic excitation of atomic force microscope cantilevers in liquids. *Applied Physics Letters* 2006, 88 (16), 163504.

[351] Dcosta, N. P.; Hoh, J. H., Calibration of optical lever sensitivity for atomic force microscopy. *Review of Scientific Instruments* 1995, 66 (10), 5096–97.

[352] Russ, J. C., Correcting image defects. In *The Image Processing Handbook* fifth edition. CRC: 2006; pp 195–268.

[353] Russ, J. C., Human vision. In *The Image Processing Handbook,* fifth edition. CRC: 2006; pp 83–134.

[354] Nguyen, C. V.; Stevens, R. M. D.; Barber, J.; Han, J.; Meyyappan, M.; Sanchez, M. I.; Larson, C.; Hinsberg, W. D., Carbon nanotube scanning probe for profiling of deep-ultraviolet and 193 nm photoresist patterns. *Applied Physics Letters* 2002, 81 (5), 901–3.

[355] Tay, A. B. H.; Thong, J. T. L., High-resolution nanowire atomic force microscope probe grown by a field-emission induced process. *Applied Physics Letters* 2004, 84 (25), 5207–9.

[356] Haycocks, J.; Jackson, K. *Detecting and addressing the surface following errors in the calibration of step heights by atomic force microscopy*, Nanoscale 2006 Seminar, Bern, Switzerland, 2006; IOP Publishing Ltd; pp 469–75.

[357] ISO, ISO 5436-1:2000. In *Geometrical Product Specifications (GPS) – Surface Texture: Profile Method; Measurement Standards – Part 1: Material Measures*, 2000.

[358] Poon, C. Y.; Bhushan, B., Comparison of surface roughness measurements by stylus profiler, AFM and non-contact optical profiler. *Wear* 1995, 190 (1), 76–88.

[359] Ramón-Torregrosa, P. J.; Rodríguez-Valverde, M. A.; Amirfazli, A.; Cabrerizo-Vílchez, M. A., Factors affecting the measurement of roughness factor of surfaces and its implications for wetting studies. *Colloids and Surfaces A: Physicochemical and Engineering Aspects* 2008, 323 (1–3), 83–93.

[360] Ohlsson, R.; Wihlborg, A.; Westberg, H. *The accuracy of fast 3D topography measurements*, 8th International Conference on Metrology and Properties of Engineering Surfaces, Huddersfield, UK, 2000; Pergamon-Elsevier Science Ltd: Oxford; pp 1899–1907.

[361] Smith, J. R.; Breakspear, S.; Campbell, S. A., AFM in surface finishing: Part II. Surface roughness. *Transactions of the Institute of Metal Finishing* 2003, 81, B55–B58.

[362] Fang, S. J.; Haplepete, S.; Chen, W.; Helms, C. R.; Edwards, H., Analyzing atomic force microscopy images using spectral methods. *Journal of Applied Physics* 1997, 82 (12), 5891–98.

[363] Thomas, T. R., Amplitude parameters. In *Rough Surfaces*. Imperial College Press: London, 1999.

[364] Gispert, M. P.; Serro, A. P.; Colaço, R.; Pires, E.; Saramago, B., Wear of ceramic coated metal-on-metal bearings used for hip replacement. *Wear* 2007, 263 (7–12), 1060–65.

[365] Patton, S. T.; Bhushan, B., Origins of friction and wear of the thin metallic layer of metal evaporated magnetic tape. *Wear* 1999, 224 (1), 126–40.

[366] Lamolle, S. F.; Monjo, M.; Lyngstadaas, S. P.; Ellingsen, J. E.; Haugen, H. J., Titanium implant surface modification by cathodic reduction in hydrofluoric acid: surface characterization and *in vivo* performance. *Journal of Biomedical Materials Research Part A* 2009, 88A (3), 581–88.

[367] Jähne, B., Segmentation. In *Practical Handbook on Image Processing for Scientific and Technical Applications*. CRC Press: 2004; pp 475–86.

[368] Klapetek, P.; Ohlídal, I.; Franta, D.; Montaigne-Ramil, A.; Bonanni, A.; Stifter, D.; Sitter, H., Atomic force microscopy characterization of ZnTe epitaxial films. *Acta Physica Slovaca* 2003, 53 (3), 223–30.

[369] Beucher, S.; Meyer, F., The morphological approach to segmentation: the watershed transformation. In *Mathematical Morphology in Image Processing* Dougherty, E. R., Ed. CRC Press: 1993; pp 433–81.

[370] El Feninat, F.; Elouatik, S.; Ellis, T. H.; Sacher, E.; Stangel, I., Quantitative assessment of surface roughness as measured by AFM: application to polished human dentin. *Applied Surface Science* 2001, 183 (3–4), 205–15.

[371] Marga, F.; Grandbois, M.; Cosgrove, D. J.; Baskin, T. I., Cell wall extension results in the coordinate separation of parallel microfibrils: evidence from scanning electron microscopy and atomic force microscopy. *Plant Journal* 2005, 43 (2), 181–90.

[372] Gan, Y., Atomic and subnanometer resolution in ambient conditions by atomic force microscopy. *Surface Science Reports* 2009, 64 (3), 99–121.

[373] Bokern, D. G.; Ducker, W. A. C.; Hunter, K. A.; McGrath, K. M., Surface imaging of a natural mineral surface using scanning-probe microscopy. *Journal of Crystal Growth* 2002, 246 (1–2), 139–49.

[374] Bowen, W. R.; Doneva, T. A., Artefacts in AFM studies of membranes: correcting pore images using fast Fourier transform filtering. *Journal of Membrane Science* 2000, 171 (1), 141–47.

[375] Vallieres, K.; Chevallier, P.; Sarra-Bournett, C.; Turgeon, S.; Laroche, G., AFM imaging of immobilized fibronectin: does the surface conjugation scheme affect the protein orientation/conformation? *Langmuir* 2007, 23 (19), 9745–51.

[376] Margeat, E.; Le Grimellec, C.; Royer, C. A., Visualization of trp repressor and its complexes with DNA by atomic force microscopy. *Biophysical Journal* 1998, 75 (6), 2712–20.

[377] Danzebrink, H. U.; Koenders, L.; Wilkening, G.; Yacoot, A.; Kunzmann, H. Advances in scanning force microscopy for dimensional metrology, *CIRP Annals – Manufacturing Technology*, 2006, 55 (2), 841–78.

[378] Taatjes, D. J.; Quinn, A. S.; Lewis, M. R.; Bovill, E. G., Quality assessment of atomic force microscopy probes by scanning electron microscopy: correlation of tip structure with rendered images. *Microscopy Research and Technique* 1999, 44 (5), 312–26.

[379] Chen, Y.; Cai, J. Y.; Liu, M. L.; Zeng, G. C.; Feng, Q.; Chen, Z. W., Research on double-probe, double- and triple-tip effects during atomic force microscopy scanning. *Scanning* 2004, 26 (4), 155–61.

[380] Nie, H.-Y.; Walzak, M. J.; Mcintyre, N. S., Use of biaxially oriented polypropylene film for evaluating and cleaning contaminated atomic force microscopy probe tips: an application to blind reconstruction. *Review of Scientific Instruments* 2002, 73 (11), 3831–36.

[381] Nowakowski, R.; Luckham, P.; Winlove, P., Imaging erythrocytes under physiological conditions by atomic force microscopy. *Biochimica et Biophysica Acta – Biomembranes* 2001, 1514 (2), 170–76.

[382] Velegol, S. B.; Pardi, S.; Li, X.; Velegol, D.; Logan, B. E., AFM imaging artifacts due to bacterial cell height and AFM tip geometry. *Langmuir* 2003, 19 (3), 851–57.

[383] Gruber, A.; Gspann, J.; Hoffmann, H., Nanostructures produced by cluster beam lithography. *Applied Physics A: Materials Science & Processing* 1999, 68 (2), 197–201.

[384] Stuart, B. H., Scanning probe microscopy. In *Analytical Techniques in Materials Conservation*, John Wiley and Sons: 2007; pp 100–03.

[385] Ploeger, R.; Murray, A.; Hesp, S.; Scalarone, D. *An investigation of the chemical changes of artists' acrylic paint films when exposed to water*, 7th Symposium on Materials Issues in Art and Archaeology, held at the 2004 MRS Fall Meeting, Boston, MA, 2004; Vandiver, P. B.; Mass, J. L.; Murray, A., Eds. Materials Research Society: pp 49–56.

[386] Steele, A.; Goddard, D.; Beech, I. B.; Tapper, R. C.; Stapleton, D.; Smith, J. R., Atomic force microscopy imaging of fragments from the Martian meteorite ALH84001. *Journal of Microscopy* 1998, 189 (1), 2–7.

[387] Nagy, K. L.; Blum, A. E., *Scanning Probe Microscopy of Clay Minerals*. Clay Minerals Society: 1994.

[388] Morris, V. J.; Gunning, A. P., Microscopy, microstructure and displacement of proteins from interfaces: implications for food quality and digestion. *Soft Matter* 2008, 4 (5), 943–51.

[389] Yang, H.; Wang, Y.; Lai, S.; An, H.; Li, Y.; Chen, F., Application of atomic force microscopy as a nanotechnology tool in food science. *Journal of Food Science* 2007, 72 (4), R65–R75.

[390] Ward, M. D., Bulk crystals to surfaces: combining X-ray diffraction and atomic force microscopy to probe the structure and formation of crystal interfaces. *Chemical Reviews* 2001, 101 (6), 1697–1725.

[391] Meyer, E.; Hug, H. J.; Bennewitz, R., Force microscopy. In *Scanning Probe Microscopy: The Lab on a Tip*. Springer: 2003; pp 45–92.

[392] Masaki, N.; Machida, K.; Kado, H.; Yokoyama, K.; Tohda, T., Molecular-resolution images of aspirin crystals with atomic force microscopy. *Ultramicroscopy* 1992, 42, 1148–54.

[393] Samori, P., Exploring supramolecular interactions and architectures by scanning force microscopies. *Chemical Society Reviews* 2005, 34 (7), 551–61.

[394] Carpick, R. W.; Salmeron, M., Scratching the surface: fundamental investigations of tribology with atomic force microscopy. *Chemical Reviews* 1997, 97 (4), 1163–94.

[395] Bhushan, B., Nanotribology and nanomechanics in nano/biotechnology. *Philosophical Transactions of the Royal Society A – Mathematical Physical and Engineering Sciences* 2008, 366 (1870), 1499–1537.

[396] Assender, H.; Bliznyuk, V.; Porfyrakis, K., How surface topography relates to materials' properties. *Science* 2002, 297 (5583), 973–76.

[397] Neogi, P., Length scales and roughness on a growing solid surface: a review. *Journal of Electroanalytical Chemistry* 2006, 595 (1), 1–10.

[398] Jandt, K. D., Atomic force microscopy of biomaterials surfaces and interfaces. *Surface Science* 2001, 491 (3), 303–32.

[399] Agnihotri, A.; Garrett, J. T.; Runt, J.; Siedlecki, C. A., Atomic force microscopy visualization of poly(urethane urea) microphase rearrangements under aqueous environment. *Journal of Biomaterials Science – Polymer Edition* 2006, 17 (1–2), 227–38.

[400] Marti, O.; Stifter, T.; Waschipky, H.; Quintus, M.; Hild, S., Scanning probe microscopy of heterogeneous polymers. *Colloids and Surfaces A: Physicochemical and Engineering Aspects* 1999, 154 (1–2), 65–73.

[401] Simon, A.; Durrieu, M. C., Strategies and results of atomic force microscopy in the study of cellular adhesion. *Micron* 2006, 37 (1), 1–13.

[402] Tak, Y.-H.; Kim, K.-B.; Park, H.-G.; Lee, K.-H.; Lee, J.-R., Criteria for ITO (indium-tin-oxide) thin film as the bottom electrode of an organic light emitting diode. *Thin Solid Films* 2002, 411 (1), 12–16.

[403] Whitehead, K. A.; Verran, J. The effect of surface topography on the retention of micro-organisms, *Food and Bioproducts Processing* 2006, 84 (4), 253–59.

[404] Mitik-Dineva, N.; Wang, J.; Truong, V. K.; Stoddart, P.; Malherbe, F.; Crawford, R. J.; Ivanova, E. P., *Escherichia coli, Pseudomonas aeruginosa*, and *Staphylococcus aureus* attachment patterns on glass surfaces with nanoscale roughness. *Current Microbiology* 2009, 58 (3), 268–73.

[405] Swerts, J.; Temst, K.; Van Bael, M. J.; Van Haesendonck, C.; Bruynseraede, Y., Magnetic domain wall trapping by in-plane surface roughness modulation. *Applied Physics Letters* 2003, 82 (8), 1239–41.

[406] Cai, K. Y.; Muller, M.; Bossert, J.; Rechtenbach, A.; Jandt, K. D., Surface structure and composition of flat titanium thin films as a function of film thickness and evaporation rate. *Applied Surface Science* 2005, 250 (1–4), 252–67.

[407] Yu, E. T., Nanoscale characterization of semiconductor materials and devices using scanning probe techniques. *Materials Science & Engineering R-Reports* 1996, 17 (4–5), 147–206.

[408] Walther, F.; Heckl, W. M.; Stark, R. W., Evaluation of nanoscale roughness measurements on a plasma treated SU-8 polymer surface by atomic force microscopy. *Applied Surface Science* 2008, 254 (22), 7290–95.

[409] Shellenberger, K.; Logan, B. E., Effect of molecular scale roughness of glass beads on colloidal and bacterial deposition. *Environmental Science & Technology* 2002, 36 (2), 184–89.

[410] Smith, J. R.; Swift, J. A., Maple syrup urine disease hair reveals the importance of 18-methyleicosanoic acid in cuticular delamination. *Micron* 2005, 36 (3), 261–66.

[411] Cacciafesta, P.; Hallam, K. R.; Watkinson, A. C.; Allen, G. C.; Miles, M. J.; Jandt, K. D., Visualisation of human plasma fibrinogen adsorbed on titanium implant surfaces with different roughness. *Surface Science* 2001, 491 (3), 405–20.

[412] Cacciafesta, P.; Hallam, K. R.; Oyedepo, C. A.; Humphris, A. D. L.; Miles, M. J.; Jandt, K. D., Characterization of ultraflat titanium oxide surfaces. *Chemistry of Materials* 2002, 14 (2), 777–89.

[413] MacDonald, D. E.; Markovic, B.; Allen, M.; Somasundaran, P.; Boskey, A. L., Surface analysis of human plasma fibronectin adsorbed to commercially pure titanium materials. *Journal of Biomedical Materials Research* 1998, 41 (1), 120–30.

[414] Larsson, C.; Thomsen, P.; Aronsson, B. O.; Rodahl, M.; Lausmaa, J.; Kasemo, B.; Ericson, L. E., Bone response to surface-modified titanium implants: studies on the early tissue response to machined and electropolished implants with different oxide thicknesses. *Biomaterials* 1996, 17 (6), 605–16.

[415] Méndez-Vilas, A.; Bruque, J. M.; Gonzalez-Martin, M. L., Sensitivity of surface roughness parameters to changes in the density of scanning points in multi-scale AFM studies. Application to a biomaterial surface. *Ultramicroscopy* 2007, 107 (8), 617–25.

[416] Macdonald, W.; Campbell, P.; Fisher, J.; Wennerberg, A., Variation in surface texture measurements. *Journal of Biomedical Materials Research B, Applied Biomaterials* 2004, 70B (2), 262–69.

[417] Hues, S. M.; Draper, C. F.; Colton, R. J. Measurement of nanomechanical properties of metals using the atomic force microscope, *Journal of Vacuum Science and Technology B* 1994, 13 (3), 2211–14.

[418] Bhushan, B.; Koinkar, V. N. Microtribological studies of doped single-crystal silicon and polysilicon films for MEMS devices, *Sensors and Actuators A: Physical*, 1996, 57(2), 91–102.

[419] Withers, J. R.; Aston, D. E., Nanomechanical measurements with AFM in the elastic limit. *Advances in Colloid and Interface Science* 2006, 120 (1–3), 57–67.

[420] Wen, J., Some mechanical properties of typical polymer-based composites. In *Physical Properties of Polymers Handbook*, Mark, J. E., Ed. Springer: 2007; pp 487–97.

[421] Bullock, S.; Johnston, E. E.; Willson, T.; Gatenholm, P.; Wynne, K. J., Surface science of a filled polydimethylsiloxane-based alkoxysilane-cured elastomer: RTV11. *Journal of Colloid and Interface Science* 1999, 210 (1), 18–36.

[422] Raghavan, D.; Gu, X.; Nguyen, T.; VanLandingham, M.; Karim, A., Mapping polymer hetereogeneity using AFM phase imaging and nanoscale indentation. *Macromolecules* 2000, 69, 2573–83.

[423] Achalla, P.; McCormick, J.; Hodge, T.; Moreland, C.; Esnault, P.; Karim, A.; Raghavan, D., Characterization of elastomeric blends by atomic force microscopy. *Journal of Polymer Science Part B – Polymer Physics* 2006, 44 (3), 492–503.

[424] Sugimoto, Y.; Namikawa, T.; Abe, M.; Morita, S., Mapping and imaging for rapid atom discrimination: a study of frequency modulation atomic force microscopy. *Applied Physics Letters* 2009, 94 (2), 023108.

[425] Giessibl, F. J., Advances in atomic force microscopy. *Reviews of Modern Physics* 2003, 75 (3), 949–83.

[426] Kuwahara, Y., Comparison of the surface structure of the tetrahedral sheets of muscovite and phlogopite by AFM. *Physics and Chemistry of Minerals* 2001, 28 (1), 1–8.

[427] Jaschke, M.; Schonherr, H.; Wolf, H.; Butt, H.-J.; Bamberg, E.; Besocke, M. K.; Ringsdorf, H., Structure of alkyl and perfluoroalkyl disulfide and azobenzenethiol monolayers on gold (111) Revealed by atomic force microscopy. *Journal of Physical Chemistry* 1996, 100 (6), 2290–2301.

[428] Hembacher, S.; Giessibl, F. J.; Mannhart, J.; Quate, C. F., Revealing the hidden atom in graphite by low-temperature atomic force microscopy. *Proceedings of the National Academy of Sciences of the United States of America* 2003, 100 (22), 12539–42.

[429] Ohnesorge, F.; Binnig, G., True atomic-resolution by atomic force microscopy through repulsive and attractive forces. *Science* 1993, 260 (5113), 1451–56.

[430] Fukuma, T., Wideband low-noise optical beam deflection sensor with photothermal excitation for liquid-environment atomic force microscopy. *Review of Scientific Instruments* 2009, 80 (2), 023707–8.

[431] Enevoldsen, G. H.; Pinto, H. P.; Foster, A. S.; Jensen, M. C. R.; Kuhnle, A.; Reichling, M.; Hofer, W. A.; Lauritsen, J. V.; Besenbacher, F., Detailed scanning probe microscopy tip models determined from simultaneous atom-resolved AFM and STM studies of the $TiO_2(110)$ surface. *Physical Review B* 2008, 78 (4), 045416.

[432] Enevoldsen, G. H.; Foster, A. S.; Christensen, M. C.; Lauritsen, J. V.; Besenbacher, F., Noncontact atomic force microscopy studies of vacancies and hydroxyls of $TiO_2(110)$: experiments and atomistic simulations. *Physical Review B* 2007, 76 (20), 205415.

[433] Maier, S.; Pfeiffer, O.; Glatzel, T.; Meyer, E.; Filleter, T.; Bennewitz, R., Asymmetry in the reciprocal epitaxy of NaCl and KBr. *Physical Review B* 2007, 75, 195408.

[434] Giessibl, F. J.; Quate, C. F., Exploring the nanoworld with atomic force microscopy. *Physics Today* 2006, 59 (12), 44–50.

[435] Giessibl, F. J.; Hembacher, S.; Bielefeldt, H.; Mannhart, J., Subatomic features on the silicon (111)-(7 × 7) surface observed by atomic force microscopy. *Science* 2000, 289 (5478), 422–25.

[436] Feenstra, R. M.; Stroscio, J. A.; Tersoff, J.; Fein, A. P., Atom-selective imaging of the GaAs (110) surface. *Physical Review Letters* 1987, 58 (12), 1192–95.

[437] Zeinalipour-Yazdi, C. D.; Pullman, D. P., A new interpretation of the scanning tunneling microscope image of graphite. *Chemical Physics* 2008, 348 (1–3), 233–36.

[438] Albrecht, T. R.; Mizes, H. A.; Nogami, J.; Park, S. I.; Quate, C. F., Observation of tilt boundaries in graphite by scanning tunneling microscopy and associated multiple tip effects. *Applied Physics Letters* 1988, 52 (5), 362–64.

[439] Ashino, M.; Schwarz, A.; Hölsher, H.; Schwarz, U. D.; Wiesendanger, R., Interpretation of the atomic scale contrast obtained on graphite and single-walled carbon nanotubes in the dynamic mode of atomic force microscopy. *Nanotechnology* 2005, 16 (3), S134–S137.

[440] Mizes, H. A.; Park, S.; Harrison, W. A., multiple-tip interpretation of anomalous scanning-tunneling-microscopy images of layered materials. *Physical Review B* 1987, 36 (8), 4491–94.

[441] Tomanek, D.; Louie, S. G.; Mamin, H. J.; Abraham, D. W.; Thomson, R. E.; Ganz, E.; Clarke, J., Theory and observation of highly asymmetric atomic-structure in scanning-tunneling-microscopy images of graphite. *Physical Review B* 1987, 35 (14), 7790–93.

[442] Szlufarska, I.; Chandross, M.; Carpick, R. W., Recent advances in single-asperity nanotribology. *Journal of Physics D – Applied Physics* 2008, 41 (12), 123001.

[443] Mate, C. M., Force microscopy studies of the molecular origins of friction and lubrication. *IBM Journal of Research and Development* 1995, 39 (6), 617–27.

[444] Zhang, Q.; Archer, L. A., Interfacial friction of surfaces grafted with one- and two-component self-assembled monolayers. *Langmuir* 2005, 21 (12), 5405–13.

[445] Martinez-Martinez, D.; Kolodziejczyk, L.; Sánchez-López, J. C.; Fernández, A., Tribological carbon-based coatings: An AFM and LFM study. *Surface Science* 2009, 603 (7), 973–79.

[446] Oncins, G.; Garcia-Manyes, S.; Sanz, F., Study of frictional properties of a phospholipid bilayer in a liquid environment with lateral force microscopy as a function of NaCl concentration. *Langmuir* 2005, 21 (16), 7373–79.

[447] Martinez-Martinez, D.; Sanchez-Lopez, J. C.; Rojas, T. C.; Fernandez, A.; Eaton, P.; Belin, M., Structural and microtribological studies of Ti-C-N based nanocomposite coatings prepared by reactive sputtering. *Thin Solid Films* 2005, 472 (1–2), 64–70.

[448] Leggett, G. J., Friction force microscopy of self-assembled monolayers: probing molecular organisation at the nanometre scale. *Analytica Chimica Acta* 2003, 479 (1), 17–38.

[449] Krausch, G.; Hipp, M.; Boeltau, M.; Marti, O.; Mlynek, J., High-resolution imaging of polymer surfaces with chemical sensitivity. *Macromolecules* 1995, 28 (1), 260–63.

[450] Cyganik, P.; Budkowski, A.; Raczkowska, J.; Postawa, Z., AFM/LFM surface studies of a ternary polymer blend cast on substrates covered by a self-assembled monolayer. *Surface Science* 2002, 507–10, 700–6.

[451] Salaita, K.; Amarnath, A.; Maspoch, D.; Higgins, T. B.; Mirkin, C. A., Spontaneous 'phase separation' of patterned binary alkanethiol mixtures. *Journal of the American Chemical Society* 2005, 127 (32), 11283–87.

[452] Hampton, J. R.; Dameron, A. A.; Weiss, P. S., Transport rates vary with deposition time in dip-pen nanolithography. *Journal of Physical Chemistry B* 2005, 109 (49), 23118–20.

[453] Gnecco, E.; Bennewitz, R.; Gyalog, T.; Loppacher, C.; Bammerlin, M.; Meyer, E.; Güntherodt, H. J., Velocity dependence of atomic friction. *Physical Review Letters* 2000, 84 (6), 1172.

[454] Brewer, N. J.; Leggett, G. J., Chemical force microscopy of mixed self-assembled mono-layers of alkanethiols on gold: evidence for phase separation. *Langmuir* 2004, 20 (10), 4109–15.

[455] Tamayo, J.; Gonzalez, L.; Gonzalez, Y.; García, R., Compositional mapping of semicon-ductor structures by friction force microscopy. *Applied Physics Letters* 1996, 68 (16), 2297–99.

[456] Hay, M. B.; Workman, R. K.; Manne, S., Mechanisms of metal ion sorption on calcite: Composition mapping by lateral force microscopy. *Langmuir* 2003, 19 (9), 3727–40.

[457] Overney, R. M.; Meyer, E.; Frommer, J.; Brodbeck, D.; Luthi, R.; Howald, L.; Guntherodt, H. J.; Fujihira, M.; Takano, H.; Gotoh, Y., Friction measurements on phase-separated thin-films with a modified atomic force microscope. *Nature* 1992, 359 (6391), 133–35.

[458] Dufrêne, Y. F.; Barger, W. R.; Green, J. B. D.; Lee, G. U., Nanometer-scale surface properties of mixed phospholipid monolayers and bilayers. *Langmuir* 1997, 13 (18), 4779–84.

[459] Tocha, E.; Schonherr, H.; Vancso, G. J., Quantitative nanotribology by AFM: a novel universal calibration platform. *Langmuir* 2006, 22 (5), 2340–50.

[460] Scandella, L.; Meyer, E.; Howald, L.; Luthi, R.; Guggisberg, M.; Gobrecht, J.; Guntherodt, H. J. Friction forces on hydrogen passivated (110) silicon and silicon dioxide studied by scanning force microscopy, *Journal of Vacuum Science and Technology B*, 1996, 14 (2), 1255–58.

[461] Beake, B. D.; Leggett, G. J.; Shipway, P. H., Frictional, adhesive and mechanical properties of polyester films probed by scanning force microscopy. *Surface and Interface Analysis* 1999, 27 (12), 1084–91.

[462] Basnar, B.; Friedbacher, G.; Brunner, H.; Vallant, T.; Mayer, U.; Hoffmann, H., Analytical evaluation of tapping mode atomic force microscopy for chemical imaging of surfaces. *Applied Surface Science* 2001, 171 (3–4), 213–25.

[463] Ciccotti, M.; George, M.; Ranieri, V.; Wondraczek, L.; Marliere, C. Dynamic condensation of water at crack tips in fused silica glass, *Journal of Non-Crystalline Solids* 2008, 354 (2–9), 564–68.

[464] Wondraczek, L.; Dittmar, A.; Oelgardt, C.; Celarie, F.; Ciccotti, M.; Marliere, C., Real-time observation of a non-equilibrium liquid condensate confined at tensile crack tips in oxide glasses. *Journal of the American Ceramic Society* 2006, 89 (2), 746–49.

[465] Dong, R.; Yu, L. Y. E., Investigation of surface changes of nanoparticles using TM-AFM phase imaging. *Environmental Science & Technology* 2003, 37 (12), 2813–19.

[466] Xu, J.; Guo, B.-H.; Zhang, Z.-M.; Zhou, J.-J.; Jiang, Y.; Yan, S.; Li, L.; Wu, Q.; Chen, G.-Q.; Schultz, J. M., Direct AFM observation of crystal twisting and organization in banded spherulites of chiral poly(3-hydroxybutyrate-co-3-hydroxyhexanoate). *Macromolecules* 2004, 37 (11), 4118–23.

[467] Bar, G.; Thomann, Y.; Brandsch, R.; Cantow, H. J.; Whangbo, M. H., Factors affecting the height and phase images in tapping mode atomic force microscopy. Study of phase-separated polymer blends of poly(ethene-co-styrene) and poly(2,6-dimethyl-1,4-phenylene oxide). *Langmuir* 1997, 13 (14), 3807–12.

[468] Bar, G.; Ganter, M.; Brandsch, R.; Delineau, L.; Whangbo, M. H., Examination of butadiene/styrene-co-butadiene rubber blends by tapping mode atomic force microscopy. Importance of the indentation depth and reduced tip-sample energy dissipation in tapping mode atomic force microscopy study of elastomers. *Langmuir* 2000, 16 (13), 5702–11.

[469] Peponi, L.; Tercjak, A.; Gutierrez, J.; Stadler, H.; Torre, L.; Kenny, J. M.; Mondragon, I., Self-assembling of SBS block copolymers as templates for conductive silver nanocomposites. *Macromolecular Materials and Engineering* 2008, 293 (7), 568–73.

[470] Holland, N. B.; Marchant, R. E., Individual plasma proteins detected on rough biomaterials by phase imaging AFM. *Journal of Biomedical Materials Research* 2000, 51 (3), 307–15.

[471] Deleu, M.; Nott, K.; Brasseur, R.; Jacques, P.; Thonart, P.; Dufrêne, Y. F., Imaging mixed lipid monolayers by dynamic atomic force microscopy. *Biochimica et Biophysica Acta – Biomembranes* 2001, 1513 (1), 55–62.

[472] Crampton, N.; Bonass, W. A.; Kirkham, J.; Thomson, N. H., Studying silane mobility on hydrated mica using ambient AFM. *Ultramicroscopy* 2006, 106 (8–9), 765–70.

[473] García, R.; Magerle, R.; Perez, R., Nanoscale compositional mapping with gentle forces. *Nature Materials* 2007, 6 (6), 405–11.

[474] Tan, S.; Sherman, R. L.; Qin, D.; Ford, W. T., Surface heterogeneity of polystyrene latex particles determined by dynamic force microscopy. *Langmuir* 2005, 21 (1), 43–49.

[475] Dorobantu, L. S.; Bhattacharjee, S.; Foght, J. M.; Gray, M. R., Atomic force microscopy measurement of heterogeneity in bacterial surface hydrophobicity. *Langmuir* 2008, 24 (9), 4944–51.

[476] Ando, T.; Uchihashi, T.; Fukuma, T., High-speed atomic force microscopy for nano-visualization of dynamic biomolecular processes. *Progress in Surface Science* 2008, 83 (7–9), 337–437.

[477] Uchihashi, T.; Ando, T.; Yamashita, H., Fast phase imaging in liquids using a rapid scan atomic force microscope. *Applied Physics Letters* 2006, 89 (21), 213112.

[478] Argaman, M.; Golan, R.; Thomson, N. H.; Hansma, H. G., Phase imaging of moving DNA molecules and DNA molecules replicated in the atomic force microscope. *Nucleic Acids Research* 1997, 25 (21), 4379–84.

[479] Kienberger, F.; Costa, L. T.; Zhu, R.; Kada, G.; Reithmayer, M.; Chtcheglova, L.; Rankl, C.; Pacheco, A. B. F.; Thalhammer, S.; Pastushenko, V.; Heckl, W. M.; Blaas, D.; Hinterdorfer, P., Dynamic force microscopy imaging of plasmid DNA and viral RNA. *Biomaterials* 2007, 28 (15), 2403–11.

[480] Hoo, C.; Starostin, N.; West, P.; Mecartney, M., A comparison of atomic force microscopy (AFM) and dynamic light scattering (DLS) methods to characterize nanoparticle size distributions. *Journal of Nanoparticle Research* 2008, 10 (Supplement 1), 89–96.

[481] Schwarz, A.; Wiesendanger, R., Magnetic sensitive force microscopy. *Nano Today* 2008, 3 (1–2), 28–39.

[482] Han, X. D.; Zhang, Z.; Wang, Z. L., Experimental nanomechanics of one-dimensional nanomaterials by in situ microscopy. *NANO: Brief Reports and Reviews* 2007, 2 (5), 249–71.

[483] Dabbousi, B. O.; Rodriguez-Viejo, J.; Mikulec, F. V.; Heine, J. R.; Mattoussi, H.; Ober, R.; Jensen, K. F.; Bawendi, M. G., (CdSe)ZnS core-shell quantum dots: synthesis and characterization of a size series of highly luminescent nanocrystallites. *Journal of Physical Chemistry B* 1997, 101 (46), 9463–75.

[484] Maye, M. M.; Luo, J.; Han, L.; Zhong, C. J., Probing Ph-tuned morphological changes in core-shell nanoparticle assembly using atomic force microscopy. *Nano Letters* 2001, 1 (10), 575–79.

[485] Eaton, P.; Doria, G.; Pereira, E.; Baptista, P. V.; Franco, R., Imaging gold nanoparticles for DNA sequence recognition in biomedical applications. *IEEE Transactions on Nanobioscience* 2007, 6, 282–88.

[486] Miura, A.; Tanaka, R.; Uraoka, Y.; Matsukawa, N.; Yamashita, I.; Fuyuki, T., The characterization of a single discrete bionanodot for memory device applications. *Nanotechnology* 2009, 20 (12), 125702.

[487] Corduneanu, O.; Diculescu, V. C.; Chiorcea-Paquim, A. M.; Oliveira-Brett, A. M., Shape-controlled palladium nanowires and nanoparticles electrodeposited on carbon electrodes. *Journal of Electroanalytical Chemistry* 2008, 624 (1–2), 97–108.

[488] Bonanni, B.; Cannistraro, S., Gold nanoparticles on modified glass surface as height calibration standard for atomic force microscopy operating in contact and tapping mode. *Journal of Nanotechnology Online* 2005, 1, 1–14.

[489] Stiger, R. M.; Gorer, S.; Craft, B.; Penner, R. M., Investigations of electrochemical silver nanocrystal growth on hydrogen-terminated silicon(100). *Langmuir* 1999, 15 (3), 790–98.

[490] Hirai, M.; Kumar, A., Wavelength tuning of surface plasmon resonance by annealing silver-copper nanoparticles. *Journal of Applied Physics* 2006, 100 (1), 014309.

[491] Raşa, M.; Philipse, A. P., Scanning probe microscopy on magnetic colloidal particles. *Journal of Magnetism and Magnetic Materials* 2002, 252 (1–3), 101–3.

[492] Sweetman, A.; Sharp, P.; Stannard, A.; Gangopadhyay, S.; Moriarty, P. J. *AFM of self-organised nanoparticle arrays: frequency modulation, amplitude modulation, and force spectroscopy*, Conference on Nanostructured Thin Films, SPIE: San Diego, CA, 2008; pp 704102–11.

[493] Gomes, I.; Santos, N. C.; Oliveira, L. M. A.; Quintas, A.; Eaton, P.; Pereira, E.; Franco, R., Probing surface properties of cytochrome c@Au bionanoconjugates. *Journal of Physical Chemistry C* 2008, 112 (42), 16340–47.

[494] Katz, E.; Willner, I., Integrated nanoparticle-biomolecule hybrid systems: synthesis, properties, and applications. *Angewandte Chemie International Edition* 2004, 43 (45), 6042–6108.

[495] Joralemon, M. J.; Murthy, K. S.; Remsen, E. E.; Becker, M. L.; Wooley, K. L., Synthesis, characterization, and bioavailability of mannosylated shell cross-linked nanoparticles. *Biomacromolecules* 2004, 5 (3), 903–13.

[496] Dukette, T. E.; Mackay, M. E.; Van Horn, B.; Wooley, K. L.; Drockenmuller, E.; Malkoch, M.; Hawker, C. J., Conformation of intramolecularly cross-linked polymer nanoparticles on solid substrates. *Nano Letters* 2005, 5 (9), 17049.

[497] Wei, Z. Q.; Mieszawska, A. J.; Zamborini, F. P., Synthesis and manipulation of high aspect ratio gold nanorods grown directly on surfaces. *Langmuir* 2004, 20 (11), 4322–26.

[498] Hsieh, S. C.; Meltzer, S.; Wang, C. R. C.; Requicha, A. A. G.; Thompson, M. E.; Koel, B. E., Imaging and manipulation of gold nanorods with an atomic force microscope. *Journal of Physical Chemistry B* 2002, 106 (2), 231–34.

[499] Fu, Y.; Ferdos, F.; Sadeghi, M.; Wang, S. M.; Larsson, A., Photoluminescence of an assembly of size-distributed self-assembled InAs quantum dots. *Journal of Applied Physics* 2002, 92 (6), 3089–92.

[500] Wang, Y.; Wei, W.; Maspoch, D.; Wu, J.; Dravid, V. P.; Mirkin, C. A., Superparamagnetic sub-5 nm Fe@C nanoparticles: isolation, structure, magnetic properties, and directed assembly. *Nano Letters* 2008, 8 (11), 3761–65.

[501] Wu, B.; Heidelberg, A.; Boland, J. J., Mechanical properties of ultrahigh-strength gold nanowires. *Nature Materials* 2005, 4 (7), 525–29.

[502] Falvo, M. R.; Clary, G. J.; Taylor, R. M.; Chi, V.; Brooks, F. P.; Washburn, S.; Superfine, R., Bending and buckling of carbon nanotubes under large strain. *Nature* 1997, 389 (6651), 582–84.

[503] Treacy, M. M. J.; Ebbesen, T. W.; Gibson, J. M., Exceptionally high Young's modulus observed for individual carbon nanotubes. *Nature* 1996, 381 (6584), 678–80.

[504] Bellucci, S., Carbon nanotubes: physics and applications. *physica status solidi (c)* 2005, 2 (1), 34–47.

[505] Yap, H. W.; Lakes, R. S.; Carpick, R. W., Mechanical instabilities of individual multiwalled carbon nanotubes under cyclic axial compression. *Nano Letters* 2007, 7 (5), 1149–54.

[506] Wong, E. W.; Sheehan, P. E.; Lieber, C. M., Nanobeam mechanics: elasticity, strength, and toughness of nanorods and nanotubes. *Science* 1997, 277 (5334), 1971–75.

[507] Song, J. H.; Wang, X. D.; Riedo, E.; Wang, Z. L., Elastic property of vertically aligned nanowires. *Nano Letters* 2005, 5 (10), 1954–58.

[508] Marszalek, P. E.; Greenleaf, W. J.; Li, H. B.; Oberhauser, A. F.; Fernandez, J. M., Atomic force microscopy captures quantized plastic deformation in gold nanowires. *Proceedings of the National Academy of Sciences of the United States of America* 2000, 97 (12), 6282–86.

[509] Tan, E. P. S.; Lim, C. T., Mechanical characterization of nanofibers – a review. *Composites Science and Technology* 2006, 66 (9), 1102–11.

[510] Zhu, Y.; Ke, C.; Espinosa, H. D., Experimental techniques for the mechanical characterization of one-dimensional nanostructures. *Experimental Mechanics* 2007, 47 (1), 7–24.

[511] Bigioni, T. P.; Cruden, B. A., Atomic force and optical microscopy characterization of the deformation of individual carbon nanotubes and nanofibers. *Journal of Nanomaterials* 2008, 352109.

[512] Yap, H. W.; Lakes, R. S.; Carpick, R. W., Negative stiffness and enhanced damping of individual multiwalled carbon nanotubes. *Physical Review B* 2008, 77 (4), 045423.

[513] Iijima, S.; Brabec, C.; Maiti, A.; Bernholc, J., Structural flexibility of carbon nanotubes. *Journal of Chemical Physics* 1996, 104 (5), 2089–92.

[514] Lee, C.; Wei, X. D.; Kysar, J. W.; Hone, J., Measurement of the elastic properties and intrinsic strength of monolayer graphene. *Science* 2008, 321 (5887), 385–8.

[515] Garcia-Sanchez, D.; van der Zande, A. M.; Paulo, A. S.; Lassagne, B.; McEuen, P. L.; Bachtold, A., Imaging mechanical vibrations in suspended graphene sheets. *Nano Letters* 2008, 8 (5), 1399–1403.

[516] Frank, I. W.; Tanenbaum, D. M.; Van der Zande, A. M.; McEuen, P. L., Mechanical properties of suspended graphene sheets. *Journal of Vacuum Science & Technology B* 2007, 25 (6), 2558–61.

[517] Edwards, S., The visionaries. In *The Nanotech Pioneers: Where Are They Taking Us?*, Wiley-VCH: 2006; pp 15–18.

[518] Luan, B.; Robbins, M. O., The breakdown of continuum models for mechanical contacts. *Nature* 2005, 435 (7044), 929–32.

[519] Li, G.; Xi, N.; Wang, D. H., *In situ* sensing and manipulation of molecules in biological samples using a nanorobotic system. *Nanomedicine: Nanotechnology, Biology and Medicine* 2005, 1 (1), 31–40.

[520] Sitti, M.; Hashimoto, H., Tele-nanorobotics using an atomic force microscope as a nanorobot and sensor. *Advanced Robotics* 1999, 13 (4), 417–36.

[521] Taylor II, R. M.; Superfine, R., Advanced interfaces to scanned-probe microscopes. In *Handbook of Nanostructured Materials and Nanotechnology*, Nalwa, H. S., Ed. Academic Press, New York, 1999; Vol. 1, pp 271–308.

[522] Ho-Yin, C.; Ning, X.; Jiangbo, Z.; Guangyong, L. *A deterministic process for fabrication and assembly of single carbon nanotube based devices*, 5th IEEE Conference on Nanotechnology 2005; pp 713–16.

[523] Carlsson, S. B.; Junno, T.; Montelius, L.; Samuelson, L., Mechanical tuning of tunnel gaps for the assembly of single-electron transistors. *Applied Physics Letters* 1999, 75 (10), 1461–63.

[524] Junno, T.; Carlsson, S. B.; Xu, H. Q.; Montelius, L.; Samuelson, L., Fabrication of quantum devices by angstrom-level manipulation of nanoparticles with an atomic force microscope. *Applied Physics Letters* 1998, 72 (5), 548–50.

[525] Harel, E.; Meltzer, S. E.; Requicha, A. A. G.; Thompson, M. E.; Koel, B. E., Fabrication of polystyrene latex nanostructures by nanomanipulation and thermal processing. *Nano Letters* 2005, 5 (12), 2624–29.

[526] Hansen, L. T.; Kuhle, A.; Sorensen, A. H.; Bohr, J.; Lindelof, P. E., A technique for positioning nanoparticles using an atomic force microscope. *Nanotechnology* 1998, 9 (4), 337–42.

[527] Meltzer, S.; Resch, R.; Koel, B. E.; Thompson, M. E.; Madhukar, A.; Requicha, A. A. G.; Will, P., Fabrication of nanostructures by hydroxylamine seeding of gold nanoparticle templates. *Langmuir* 2001, 17 (5), 1713–18.

[528] Resch, R.; Baur, C.; Bugacov, A.; Koel, B. E.; Madhukar, A.; Requicha, A. A. G.; Will, P., Building and manipulating three-dimensional and linked two- dimensional structures of nanoparticles using scanning force microscopy. *Langmuir* 1998, 14 (23), 6613–16.

[529] Xie, H.; Haliyo, D. S.; Regnier, S., Parallel imaging/manipulation force microscopy. *Applied Physics Letters* 2009, 94 (15), 153106–3.

[530] Palacio, M.; Bhushan, B., A nanoscale friction investigation during the manipulation of nanoparticles in controlled environments. *Nanotechnology* 2008, 19 (31), 315710.

[531] Tranvouez, E.; Orieux, A.; Boer-Duchemin, E.; Devillers, C. H.; Huc, V.; Comtet, G.; Dujardin, G., Manipulation of cadmium selenide nanorods with an atomic force microscope. *Nanotechnology* 2009, 20 (16), 165304.

[532] Rubio-Sierra, F. J.; Heckl, W. M.; Stark, R. W., Nanomanipulation by atomic force microscopy. *Advanced Engineering Materials* 2005, 7 (4), 193–96.

[533] Stark, R. W.; Rubio-Sierra, F. J.; Thalhammer, S.; Heckl, W. M., Combined nanomanipulation by atomic force microscopy and UV-laser ablation for chromosomal dissection. *European Biophysics Journal* 2003, 32 (1), 33–9.

[534] Baptista, P.; Pereira, E.; Eaton, P.; Doria, G.; Miranda, A.; Gomes, I.; Quaresma, P.; Franco, R., Gold nanoparticles for the development of clinical diagnosis methods. *Analytical and Bioanalytical Chemistry* 2008, 391 (3), 943–50.

[535] Biju, V.; Itoh, T.; Anas, A.; Sujith, A.; Ishikawa, M., Semiconductor quantum dots and metal nanoparticles: syntheses, optical properties, and biological applications. *Analytical and Bioanalytical Chemistry* 2008, 391 (7), 2469–95.

[536] Pankhurst, Q. A.; Connolly, J.; Jones, S. K.; Dobson, J., Applications of magnetic nanoparticles in biomedicine. *Journal of Physics D – Applied Physics* 2003, 36 (13), R167–R181.

[537] Sperling, R. A.; Rivera Gil, P.; Zhang, F.; Zanella, M.; Parak, W. J., Biological applications of gold nanoparticles. *Chemical Society Reviews* 2008, 37 (9), 1896–1908.

[538] Pellegrino, T.; Kudera, S.; Liedl, T.; Javier, A. M.; Manna, L.; Parak, W. J., On the development of colloidal nanoparticles towards multifunctional structures and their possible use for biological applications. *Small* 2005, 1 (1), 48–63.

[539] Wei, G.; Wang, L.; Zhou, H. L.; Liu, Z. G.; Song, Y. H.; Li, Z. A., Electrostatic assembly of CTAB-capped silver nanoparticles along predefined gimel-DNA template. *Applied Surface Science* 2005, 252 (5), 1189–96.

[540] Nakao, H.; Shiigi, H.; Yamamoto, Y.; Tokonami, S.; Nagaoka, T.; Sugiyama, S.; Ohtani, T., Highly ordered assemblies of Au nanoparticles organized on DNA. *Nano Letters* 2003, 3 (10), 1391–94.

[541] Jaganathan, H.; Ivanisevic, A., Heterostructured DNA templates: a combined magnetic force microscopy and circular dichroism study. *Applied Physics Letters* 2008, 93 (26), 263104–3.

[542] Braun, E.; Eichen, Y.; Sivan, U.; Ben-Yoseph, G., DNA-templated assembly and electrode attachment of a conducting silver wire. *Nature* 1998, 391 (6669), 775–78.

[543] Gu, Q.; Cheng, C. D.; Suryanarayanan, S.; Dai, K.; Haynie, D. T., DNA-templated fabrication of nickel nanocluster chains. *Physica E – Low-Dimensional Systems & Nanostructures* 2006, 33 (1), 92–98.

[544] Sun, L. L.; Sun, Y. J.; Xu, F. G.; Zhang, Y.; Yang, T.; Guo, C. L.; Liu, Z. L.; Li, Z., Atomic force microscopy and surface-enhanced Raman scattering detection of DNA based on DNA-nanoparticle complexes. *Nanotechnology* 2009, 20 (12), 125502.

[545] Ebenstein, Y.; Gassman, N.; Kim, S.; Antelman, J.; Kim, Y.; Ho, S.; Samuel, R.; Michalet, X.; Weiss, S., Lighting up individual DNA binding proteins with quantum dots. *Nano Letters* 2009, 9 (4), 1598–1603.

[546] Patolsky, F.; Weizmann, Y.; Lioubashevski, O.; Willner, I., Au-nanoparticle nanowires based on DNA and polylysine templates. *Angewandte Chemie International Edition* 2002, 41 (13), 2323–27.

[547] Braun, G.; Inagaki, K.; Estabrook, R. A.; Wood, D. K.; Levy, E.; Cleland, A. N.; Strouse, G. F.; Reich, N. O., Gold nanoparticle decoration of DNA on silicon. *Langmuir* 2005, 21 (23), 10699–701.

[548] Zhao, W.; Brook, M. A.; Li, Y. F., Design of gold nanoparticle-based colorimetric biosensing assays. *ChemBioChem* 2008, 9 (15), 2363–71.

[549] Braun, G.; Diechtierow, M.; Wilkinson, S.; Schmidt, F.; Hüben, M.; Weinhold, E.; Reich, N. O., Enzyme-directed positioning of nanoparticles on large DNA templates. *Bioconjugate Chemistry* 2008, 19 (2), 476–79.

[550] Zdrojek, M.; Melin, T.; Diesinger, H.; Stievenard, D.; Gebicki, W.; Adamowicz, L., Charging and discharging processes of carbon nanotubes probed by electrostatic force microscopy. *Journal of Applied Physics* 2006, 100 (11), 114326.

[551] Yaish, Y.; Park, J. Y.; Rosenblatt, S.; Sazonova, V.; Brink, M.; McEuen, P. L., Electrical nanoprobing of semiconducting carbon nanotubes using an atomic force microscope. *Physical Review Letters* 2004, 92 (4), 046401.

[552] Bachtold, A.; Fuhrer, M. S.; Plyasunov, S.; Forero, M.; Anderson, E. H.; Zettl, A.; McEuen, P. L., Scanned probe microscopy of electronic transport in carbon nanotubes. *Physical Review Letters* 2000, 84 (26), 6082–85.

[553] Park, J. Y., Electrically tunable defects in metallic single-walled carbon nanotubes. *Applied Physics Letters* 2007, 90 (2), 023112.

[554] Lu, W.; Wang, D.; Chen, L., Near-static dielectric polarization of individual carbon nanotubes. *Nano Letters* 2007, 7 (9), 2729–33.

[555] Park, J. Y.; Yaish, Y.; Brink, M.; Rosenblatt, S.; McEuen, P. L., Electrical cutting and nicking of carbon nanotubes using an atomic force microscope. *Applied Physics Letters* 2002, 80 (23), 4446–48.

[556] Bozovic, D.; Bockrath, M.; Hafner, J. H.; Lieber, C. M.; Park, H.; Tinkham, M., Electronic properties of mechanically induced kinks in single-walled carbon nanotubes. *Applied Physics Letters* 2001, 78 (23), 3693–95.

[557] Bozovic, D.; Bockrath, M.; Hafner, J. H.; Lieber, C. M.; Park, H.; Tinkham, M., Plastic deformations in mechanically strained single-walled carbon nanotubes. *Physical Review B* 2003, 67 (3), 033407.

[558] Barboza, A. P. M.; Gomes, A. P.; Archanjo, B. S.; Araujo, P. T.; Jorio, A.; Ferlauto, A. S.; Mazzoni, M. S. C.; Chacham, H.; Neves, B. R. A., Deformation induced semiconductor-metal transition in single wall carbon nanotubes probed by electric force microscopy. *Physical Review Letters* 2008, 100 (25), 256804.

[559] Woodside, M. T.; McEuen, P. L., Scanned probe imaging of single-electron charge states in nanotube quantum dots. *Science* 2002, 296 (5570), 1098–1101.

[560] Tanaka, I.; Kamiya, I.; Sakaki, H.; Qureshi, N.; Allen, J. S. J.; Petroff, P. M., Imaging and probing electronic properties of self-assembled InAs quantum dots by atomic force microscopy with conductive tip. *Applied Physics Letters* 1999, 74 (6), 844–46.

[561] Birjukovs, P.; Petkov, N.; Xu, J.; Svirksts, J.; Boland, J. J.; Holmes, J. D.; Erts, D., Electrical characterization of bismuth sulfide nanowire arrays by conductive atomic force microscopy. *Journal of Physical Chemistry C* 2008, 112 (49), 19680–85.

[562] Pan, N.; Wang, X. P.; Zhang, K.; Hu, H. L.; Xu, B.; Li, F. Q.; Hou, J. G., An approach to control the tip shapes and properties of ZnO nanorods. *Nanotechnology* 2005, 16 (8), 1069–72.

[563] He, J. H.; Ho, S. T.; Wu, T. B.; Chen, L. J.; Wang, Z. L., Electrical and photoelectrical performances of nano-photodiode based on ZnO nanowires. *Chemical Physics Letters* 2007, 435 (1–3), 119–22.

[564] Lucchesi, M.; Privitera, G.; Labardi, M.; Prevosto, D.; Capaccioli, S.; Pingue, P., Electrostatic force microscopy and potentiometry of realistic nanostructured systems. *Journal of Applied Physics* 2009, 105 (5), 54301.

[565] Erts, D.; Polyakov, B.; Dalyt, B.; Morris, M. A.; Ellingboe, S.; Boland, J.; Holmes, J. D., High density germanium nanowire assemblies: contact challenges and electrical characterization. *Journal of Physical Chemistry B* 2006, 110 (2), 820–26.

[566] Kalinin, S. V.; Shin, J.; Jesse, S.; Geohegan, D.; Baddorf, A. P.; Lilach, Y.; Moskovits, M.; Kolmakov, A., Electronic transport imaging in a multiwire SnO2 chemical field-effect transistor device. *Journal of Applied Physics* 2005, 98 (4), 044503.

[567] Xu, D. G.; Watt, G. D.; Harb, J. N.; Davis, R. C., Electrical conductivity of ferritin proteins by conductive AFM. *Nano Letters* 2005, 5 (4), 571–77.

[568] Axford, D. N.; Davis, J. J., Electron flux through apo-and holoferritin. *Nanotechnology* 2007, 18 (14), 145502.

[569] Zhao, J. W.; Davis, J. J., Molecular electron transfer of protein junctions characterised by conducting atomic force microscopy. *Colloids and Surfaces B: Biointerfaces* 2005, 40 (3–4), 189–94.

[570] MacCuspie, R. I.; Nuraje, N.; Lee, S.-Y.; Runge, A.; Matsui, H., Comparison of electrical properties of viruses studied by AC capacitance scanning probe microscopy. *Journal of the American Chemical Society* 2008, 130 (3), 887–91.

[571] Luo, E. Z.; Wilson, I. H.; Yan, X.; Xu, J. B., Probing electron conduction at the microscopic level in percolating nanocomposites by conducting atomic-force microscopy. *Physical Review B* 1998, 57 (24), R15120.

[572] Olbrich, A.; Ebersberger, B.; Boit, C.; Vancea, J.; Hoffmann, H., A new AFM-based tool for testing dielectric quality and reliability on a nanometer scale. *Microelectronics Reliability* 1999, 39 (6–7), 941–46.

[573] Aliev, A. E.; Oh, J.; Kozlov, M. E.; Kuznetsov, A. A.; Fang, S.; Fonseca, A. F.; Ovalle, R.; Lima, M. D.; Haque, M. H.; Gartstein, Y. N.; Zhang, M.; Zakhidov, A. A.; Baughman, R. H., Giant-stroke, superelastic carbon nanotube aerogel muscles. *Science* 2009, 323 (5921), 1575–78.

[574] Zdrojek, M.; Melin, T.; Boyaval, C.; Stievenard, D.; Jouault, B.; Wozniak, M.; Huczko, A.; Gebicki, W.; Adamowicz, L., Charging and emission effects of multiwalled carbon nanotubes probed by electric force microscopy. *Applied Physics Letters* 2005, 86 (21), 213114.

[575] Vitali, L.; Burghard, M.; Wahl, P.; Schneider, M. A.; Kern, K., Local pressure-induced metallization of a semiconducting carbon nanotube in a crossed junction. *Physical Review Letters* 2006, 96 (8), 086804.

[576] Hansma, P. K.; Cleveland, J. P.; Radmacher, M.; Walters, D. A.; Hillner, P. E.; Bezanilla, M.; Fritz, M.; Vie, D.; Hansma, H. G.; Prater, C. B.; Massie, J.; Fukunaga, L.; Gurley, J.; Elings, V., Tapping mode atomic-force microscopy in liquids. *Applied Physics Letters* 1994, 64 (13), 1738–40.

[577] Morris, V. J.; Kirby, A. R.; Gunning, A. P., *Atomic Force Microscopy for Biologists*. Imperial College Press: London, 1999.

[578] Jena, B. P.; Hörber, J. K. H., *Atomic Force Microscopy in Cell Biology*. Academic Press: San Diego, 2002; Vol. 68.

[579] Parot, P.; Dufrêne, Y. F.; Hinterdorfer, P.; Grimellec, C. L.; Navajas, D.; Pellequer, J.-L.; Scheuring, S., Past, present and future of atomic force microscopy in life sciences and medicine. *Journal of Molecular Recognition* 2007, 20 (6), 418–31.

[580] Fuss, M.; Luna, M.; Alcantara, D.; de la Fuente, J. M.; Enriquez-Navas, P. M.; Angulo, J.; Penades, S.; Briones, F., Carbohydrate–carbohydrate interaction prominence in 3D supramolecular self-assembly. *Journal of Physical Chemistry B* 2008, 112 (37), 11595–600.

[581] De la Fuente, J. M.; Alcantara, D.; Eaton, P.; Crespo, P.; Rojas, T. C.; Fernandez, A.; Hernando, A.; Penades, S., Gold and gold-iron oxide magnetic glyconanoparticles: synthesis, characterization and magnetic properties. *Journal of Physical Chemistry B* 2006, 110 (26), 13021–28.

[582] Kirby, A. R.; Gunning, A. P.; Morris, V. J., Imaging polysaccharides by atomic force microscopy. *Biopolymers* 1996, 38 (3), 355–66.

[583] Adams, E. L.; Kroon, P. A.; Williamson, G.; Morris, V. J., Characterisation of heterogeneous arabinoxylans by direct imaging of individual molecules by atomic force microscopy. *Carbohydrate Research* 2003, 338 (8), 771–80.

[584] Misevic, G. N., Atomic force microscopy measurements – measurements of binding strength between a single pair of molecules in physiological solutions. *Molecular Biotechnology* 2001, 18 (2), 149–53.

[585] Stolz, M.; Stoffler, D.; Aebi, U.; Goldsbury, C., Monitoring biomolecular interactions by time-lapse atomic force microscopy. *Journal of Structural Biology* 2000, 131 (3), 171–80.

[586] Forman, J. R.; Clarke, J., Mechanical unfolding of proteins: insights into biology, structure and folding. *Current Opinion in Structural Biology* 2007, 17 (1), 58–66.

[587] Zhao, J. W.; Davis, J. J.; Sansom, M. S. P.; Hung, A., Exploring the electronic and mechanical properties of protein using conducting atomic force microscopy. *Journal of the American Chemical Society* 2004, 126 (17), 5601–9.

[588] Vinckier, A.; Gervasoni, P.; Zaugg, F.; Ziegler, U.; Lindner, P.; Groscurth, P.; Pluckthun, A.; Semenza, G., Atomic force microscopy detects changes in the interaction forces between GroEL and substrate proteins. *Biophysical Journal* 1998, 74 (6), 3256–63.

[589] Krishna, K. A.; Rao, G. V.; Rao, K., Chaperonin GroEL: structure and reaction cycle. *Current Protein and Peptide Science* 2007, 8 (5), 418–25.

[590] Mou, J. X.; Sheng, S. T.; Ho, R. Y.; Shao, Z. F., Chaperonins GroEL and GroES: views from atomic force microscopy. *Biophysical Journal* 1996, 71 (4), 2213–21.

[591] Leung, C.; Palmer, R. E., Adsorption of a model protein, the GroEL chaperonin, on surfaces. *Journal of Physics – Condensed Matter* 2008, 20 (35), 353001.

[592] Valle, F.; DeRose, J. A.; Dietler, G.; Kawe, M.; Pluckthun, A.; Semenza, G., AFM structural study of the molecular chaperone GroEL and its two-dimensional crystals: an ideal 'living' calibration sample. *Ultramicroscopy* 2002, 93 (1), 83–89.

[593] Yokokawa, M.; Wada, C.; Ando, T.; Sakai, N.; Yagi, A.; Yoshimura, S. H.; Takeyasu, K., Fast-scanning atomic force microscopy reveals the ATP/ADP-dependent conformational changes of GroEL. *EMBO Journal* 2006, 25 (19), 4567–76.

[594] Sit, P. S.; Marchant, R. E., Surface-dependent differences in fibrin assembly visualized by atomic force microscopy, *Surface Science* 2001, 491 (3), 421–32.

[595] Baselt, D. R.; Revel, J. P.; Baldeschwieler, J. D., Subfibrillar structure of type-I collagen observed by atomic-force microscopy. *Biophysical Journal* 1993, 65 (6), 2644–55.

[596] Gale, M.; Pollanen, M. S.; Markiewicz, P.; Goh, M. C., Sequential assembly of collagen revealed by atomic-force microscopy. *Biophysical Journal* 1995, 68 (5), 2124–8.

[597] Paige, M. F.; Rainey, J. K.; Goh, M. C., A study of fibrous long spacing collagen ultrastructure and assembly by atomic force microscopy. *Micron* 2001, 32 (3), 341–53.

[598] Abraham, L. C.; Zuena, E.; Perez-Ramirez, B.; Kaplan, D. L., Guide to collagen characterization for biomaterial studies. *Journal of Biomedical Materials Research B, Applied Biomaterials* 2008, 87 (1), 264–85.

[599] Revenko, I.; Sommer, F.; Minh, D. T.; Garrone, R.; Franc, J. M., Atomic-force microscopy study of the collagen fiber structure. *Biology of the Cell* 1994, 80 (1), 67–9.

[600] Fotiadis, D.; Scheuring, S.; Müller, S. A.; Engel, A.; Müller, D. J., Imaging and manipulation of biological structures with the AFM. *Micron* 2002, 33 (4), 385–97.

[601] Anselmetti, D.; Luthi, R.; Meyer, E.; Richmond, T.; Dreier, M.; Frommer, J. E.; Guntherodt, H. J., Attractive-mode imaging of biological materials with dynamic force microscopy. *Nanotechnology* 1994, 5 (2), 87–94.

[602] Hansma, H. G.; Revenko, I.; Kim, K.; Laney, D. E., Atomic force microscopy of long and short double-stranded, single-stranded and triple-stranded nucleic acids. *Nucleic Acids Research* 1996, 24 (4), 713–20.

[603] Palacios-Lidón, E.; Pérez-García, B.; Colchero, J., Enhancing dynamic scanning force microscopy in air: as close as possible. *Nanotechnology* 2009, 20 (8), 085707.

[604] Giro, A.; Bergia, A.; Zuccheri, G.; Bink, H. H. J.; Pleij, C. W. A.; Samori, B., Single molecule studies of RNA secondary structure: AFM of TYMV viral RNA. *Microscopy Research and Technique* 2004, 65 (4–5), 235–45.

[605] Bonin, M.; Oberstrass, J.; Lukacs, N.; Ewert, K.; Oesterschulze, E.; Kassing, R.; Nellen, W., Determination of preferential binding sites for anti-dsRNA antibodies on double-stranded RNA by scanning force microscopy. *RNA – Publication of the RNA Society* 2000, 6 (4), 563–70.

[606] Bonin, M.; Oberstrass, J.; Vogt, U.; Wassenegger, M.; Nellen, W., Binding of IRE-BP to its cognate RNA sequence: SFM studies on a universal RNA backbone for the analysis of RNA-protein interaction. *Biological Chemistry* 2001, 382 (8), 1157–62.

[607] Asami, Y.; Murakami, M.; Shimizu, M.; Pisani, F. M.; Hayata, I.; Nohmi, T., Visualization of the interaction between archaeal DNA polymerase and uracil-containing DNA by atomic force microscopy. *Genes to Cells* 2006, 11 (1), 3–11.

[608] Kasas, S.; Thomson, N. H.; Smith, B. L.; Hansma, H. G.; Zhu, X. S.; Guthold, M.; Bustamante, C.; Kool, E. T.; Kashlev, M.; Hansma, P. K., *Escherichia coli* RNA polymerase activity observed using atomic force microscopy. *Biochemistry* 1997, 36 (3), 461–8.

[609] Engel, A.; Müller, D. J., Observing single biomolecules at work with the atomic force microscope. *Nature Structural Biology* 2000, 7 (9), 715–18.

[610] Thomson, N. H., Atomic force microscopy of DNA structure and function. In *Applied Scanning Probe Methods, Vol VI: Characterization*, Bhushan, B.; Fuchs, H.; Hosaka, S., Eds. Springer-Verlag: Berlin, 2006.

[611] Gaboriaud, F.; Dufrêne, Y. F., Atomic force microscopy of microbial cells: application to nanomechanical properties, surface forces and molecular recognition forces. *Colloids and Surfaces B: Biointerfaces* 2007, 54 (1), 10–19.

[612] Schmid, T.; Burkhard, J.; Yeo, B. S.; Zhang, W. H.; Zenobi, R., Towards chemical analysis of nanostructures in biofilms I: imaging of biological nanostructures. *Analytical and Bioanalytical Chemistry* 2008, 391 (5), 1899–1905.

[613] Zhao, L. M.; Schaefer, D.; Marten, M. R., Assessment of elasticity and topography of *Aspergillus nidulans* spores *via* atomic force microscopy. *Applied and Environmental Microbiology* 2005, 71 (2), 955–60.

[614] Plomp, M.; Leighton, T. J.; Wheeler, K. E.; Hill, H. D.; Malkin, A. J., *In vitro* high-resolution structural dynamics of single germinating bacterial spores. *Proceedings of the National Academy of Sciences of the United States of America* 2007, 104 (23), 9644–9.

[615] Plomp, M.; Malkin, A. J., Mapping of proteomic composition on the surfaces of *Bacillus* spores by atomic force microscopy-based immunolabeling. *Langmuir* 2009, 25 (1), 403–9.

[616] Das, S. K.; Das, A. R.; Guha, A. K., Adsorption behavior of mercury on functionalized *Aspergillus versicolor* mycelia: an atomic force microscopic study. *Langmuir* 2009, 25 (1), 360–66.

[617] Ma, H.; Snook, L. A.; Tian, C.; Kaminskyj, S. G. W.; Dahms, T. E. S., Fungal surface remodelling visualized by atomic force microscopy. *Mycological Research* 2006, 110 (8), 879–86.

[618] Bui, V. C.; Kim, Y. U.; Choi, S. S., Physical characteristics of *Saccharomyces cerevisiae*. *Surface and Interface Analysis* 2008, 40 (10), 1323–27.

[619] Schmatulla, A.; Maghelli, N.; Marti, O., Micromechanical properties of tobacco mosaic viruses. *Journal of Microscopy* 2007, 225 (3), 264–68.

[620] Francius, G.; Tesson, B.; Dague, E.; Martin-Jézéquel, V.; Dufrêne, Y. F., Nanostructure and nanomechanics of live *Phaeodactylum tricornutum* morphotypes. *Environmental Microbiology* 2008, 10 (5), 1344–56.

[621] Bolshakova, A. V.; Kiselyova, O. I.; Filonov, A. S.; Frolova, O. Y.; Lyubchenko, Y. L.; Yaminsky, I. V., Comparative studies of bacteria with an atomic force microscope operating in different modes. *Ultramicroscopy* 2001, 86 (1–2), 121–28.

[622] Méndez-Vilas, A.; Gallardo-Moreno, A. M.; Calzado-Montero, R.; González-Martín, M. L., AFM probing in aqueous environment of *Staphylococcus epidermidis* cells naturally immobilised on glass: physico-chemistry behind the successful immobilisation. *Colloids and Surfaces B: Biointerfaces* 2008, 63 (1), 101–9.

[623] Cross, S. E.; Kreth, J.; Zhu, L.; Qi, F. X.; Pelling, A. E.; Shi, W. Y.; Gimzewski, J. K., Atomic force microscopy study of the structure-function relationships of the biofilm-forming bacterium *Streptococcus mutans*. *Nanotechnology* 2006, 17 (4), S1–S7.

[624] Braga, P. C.; Ricci, D., Differences in the susceptibility of *Streptococcus pyogenes* to rokitamycin and erythromycin A revealed by morphostructural atomic force microscopy. *Journal of Antimicrobial Chemotherapy* 2002, 50 (4), 457–60.

[625] Suo, Z.; Yang, X.; Avci, R.; Kellerman, L.; Pascual, D. W.; Fries, M.; Steele, A., HEPES-stabilized encapsulation of *Salmonella typhimurium*. *Langmuir* 2007, 23 (3), 1365–74.

[626] Jonas, K.; Tomenius, H.; Kader, A.; Normark, S.; Romling, U.; Belova, L.; Melefors, O., Roles of curli, cellulose and BapA in *Salmonella* biofilm morphology studied by atomic force microscopy. *BMC Microbiology* 2007, 7, 70.

[627] Martinez, J. L.; Fajardo, A.; Garmendia, L.; Hernandez, A.; Linares, J. F.; Martínez-Solano, L.; Sánchez, M. B., A global view of antibiotic resistance. *FEMS Microbiology Reviews* 2009, 33 (1), 44–65.

[628] Braga, P. C.; Ricci, D., Atomic force microscopy: application to investigation of *Escherichia coli* morphology before and after exposure to cefodizime. *Antimicrobial Agents and Chemotherapy* 1998, 42 (1), 18–22.

[629] Mortensen, N. P.; Fowlkes, J. D.; Sullivan, C. J.; Allison, D. P.; Larsen, N. B.; Molin, S.; Doktycz, M. J., Effects of colistin on surface ultrastructure and nanomechanics of *Pseudomonas aeruginosa* cells. *Langmuir* 2009, 25 (6), 3728–33.

[630] Meincken, M.; Holroyd, D. L.; Rautenbach, M., Atomic force microscopy study of the effect of antimicrobial peptides on the cell envelope of *Escherichia coli*. *Antimicrobial Agents and Chemotherapy* 2005, 49 (10), 4085–92.

[631] Boyle-Vavra, S.; Hahm, J.; Sibener, S. J.; Daum, R. S., Structural and topological differences between a glycopeptide-intermediate clinical strain and glycopeptide-susceptible strains of *Staphylococcus aureus* revealed by atomic force microscopy. *Antimicrobial Agents and Chemotherapy* 2000, 44 (12), 3456–60.

[632] Francius, G.; Domenech, O.; Mingeot-Leclercq, M. P.; Dufrêne, Y. F., Direct observation of *Staphylococcus aureus* cell wall digestion by lysostaphin. *Journal of Bacteriology* 2008, 190 (24), 7904–9.

[633] Qin, Z.; Zhang, J.; Hu, Y.; Chi, Q.; Mortensen, N. P.; Qu, D.; Molin, S.; Ulstrup, J., Organic compounds inhibiting *S. epidermidis* adhesion and biofilm formation. *Ultramicroscopy* 2009, 109 (8), 881–88.

[634] Deupree, S. M.; Schoenfisch, M. H., Morphological analysis of the antimicrobial action of nitric oxide on Gram-negative pathogens using atomic force microscopy. *Acta Biomaterialia* 2009, 5 (5), 1405–15.

[635] Beckmann, M. A.; Venkataraman, S.; Doktycz, M. J.; Nataro, J. P.; Sullivan, C. J.; Morrell-Falvey, J. L.; Allison, D. P., Measuring cell surface elasticity on enteroaggregative *Escherichia coli* wild type and dispersin mutant by AFM. *Ultramicroscopy* 2006, 106 (8–9), 695–702.

[636] Schaer-Zammaretti, P.; Ubbink, J., Imaging of lactic acid bacteria with AFM-elasticity and adhesion maps and their relationship to biological and structural data. *Ultramicroscopy* 2003, 97 (1–4), 199–208.

[637] Wright, C. J.; Armstrong, I., The application of atomic force microscopy force measurements to the characterisation of microbial surfaces. *Surface and Interface Analysis* 2006, 38 (11), 1419–28.

[638] Dague, E.; Alsteens, D.; Latge, J.-P.; Verbelen, C.; Raze, D.; Baulard, A. R.; Dufrêne, Y. F., Chemical force microscopy of single live cells. *Nano Letters* 2007, 7 (10), 3026–30.

[639] Francius, G.; Lebeer, S.; Alsteens, D.; Wildling, L.; Gruber, H. J.; Hols, P.; De Keersmaecker, S.; Vanderleyden, J.; Dufrêne, Y. F., Detection, localization and conformational analysis of single polysaccharide molecules on live bacteria. *ACS Nano* 2008, 2 (9), 1921–29.

[640] Deupree, S. M.; Schoenfisch, M. H., Quantitative method for determining the lateral strength of bacterial adhesion and application for characterizing adhesion kinetics. *Langmuir* 2008, 24 (9), 47007.

[641] Boyd, R. D.; Verran, J.; Jones, M. V.; Bhakoo, M., Use of the atomic force microscope to determine the effect of substratum surface topography on bacterial adhesion. *Langmuir* 2002, 18 (6), 2343–46.

[642] Lower, S. K.; Hochella, M. F., Jr.; Beveridge, T. J., Bacterial recognition of mineral surfaces: nanoscale interactions between *Shewanella* and α-FeOOH. *Science* 2001, 292 (5520), 1360–63.

[643] Razatos, A.; Ong, Y. L.; Sharma, M. M.; Georgiou, G., Evaluating the interaction of bacteria with biomaterials using atomic force microscopy. *Journal of Biomaterials Science – Polymer Edition* 1998, 9 (12), 1361–73.

[644] Ong, Y. L.; Razatos, A.; Georgiou, G.; Sharma, M. M., Adhesion forces between *E. coli* bacteria and biomaterial surfaces. *Langmuir* 1999, 15 (8), 2719–25.

[645] Camesano, T. A.; Liu, Y.; Datta, M., Measuring bacterial adhesion at environmental interfaces with single-cell and single-molecule techniques. *Advances in Water Resources* 2007, 30 (6–7), 1470–91.

[646] Jass, J.; Tjarnhage, T.; Puu, G., From liposomes to supported, planar bilayer structures on hydrophilic and hydrophobic surfaces: an atomic force microscopy study. *Biophysical Journal* 2000, 79 (6), 3153–63.

[647] Jeuken, L. J. C.; Connell, S. D.; Henderson, P. J. F.; Gennis, R. B.; Evans, S. D.; Bushby, R. J., Redox enzymes in tethered membranes. *Journal of the American Chemical Society* 2006, 128 (5), 1711–16.

[648] Hui, S. W.; Viswanathan, R.; Zasadzinski, J. A.; Israelachvili, J. N., The structure and stability of phospholipid bilayers by atomic force microscopy. *Biophysical Journal* 1995, 68 (1), 171–78.

[649] Egawa, H.; Furusawa, K., Liposome adhesion on mica surface studied by atomic force microscopy. *Langmuir* 1999, 15 (5), 16606.

[650] Ohler, B.; Revenko, I.; Husted, C., Atomic force microscopy of nonhydroxy-galactocerebroside nanotubes and their self-assembly at the air-water interface, with applications to myelin. *Journal of Structural Biology* 2001, 133 (1), 1–9.

[651] Simons, K.; Ikonen, E., Functional rafts in cell membranes. *Nature* 1997, 387 (6633), 569–72.

[652] Dufrêne, Y. F.; Lee, G. U., Advances in the characterization of supported lipid films with the atomic force microscope. *Biochimica et Biophysica Acta – Biomembranes* 2000, 1509 (1–2), 14–41.

[653] Eeman, M.; Deleu, M.; Paquot, M.; Thonart, P.; Dufrêne, Y. F., Nanoscale properties of mixed fengycin/ceramide monolayers explored using atomic force microscopy. *Langmuir* 2005, 21 (6), 2505–11.

[654] Schneider, J.; Dufrêne, Y. F.; Barger Jr, W. R.; Lee, G. U., Atomic force microscope image contrast mechanisms on supported lipid bilayers. *Biophysical Journal* 2000, 79 (2), 1107–18.

[655] Nicolini, C.; Baranski, J.; Schlummer, S.; Palomo, J.; Lumbierres-Burgues, M.; Kahms, M.; Kuhlmann, J.; Sanchez, S.; Gratton, E.; Waldmann, H.; Winter, R., Visualizing association of N-Ras in lipid microdomains: influence of domain structure and interfacial adsorption. *Journal of the American Chemical Society* 2006, 128 (1), 192–201.

[656] Butt, H.-J.; Franz, V., Rupture of molecular thin films observed in atomic force microscopy. I. Theory. *Physical Review E* 2002, 66 (3), 031601.

[657] Sanderson, J. M., Peptide lipid interactions: insights and perspectives. *Organic & Biomolecular Chemistry* 2005, 3 (2), 201–12.

[658] El Kirat, K.; Lins, L.; Brasseur, R.; Dufrêne, Y. F., Fusogenic tilted peptides induce nanoscale holes in supported phosphatidylcholine bilayers. *Langmuir* 2005, 21 (7), 3116–21.

[659] García-Sáez, A. J.; Chiantia, S.; Salgado, J.; Schwille, P., Pore formation by a Bax-derived peptide: effect on the line tension of the membrane probed by AFM. *Biophysical Journal* 2007, 93 (1), 103–12.

[660] El Kirat, K.; Dufrêne, Y. F.; Lins, L.; Brasseur, R., The SIV tilted peptide induces cylindrical reverse micelles in supported lipid bilayers. *Biochemistry* 2006, 45 (30), 9336–41.

[661] Grandbois, M.; Clausen-Schaumann, H.; Gaub, H., Atomic force microscope imaging of phospholipid bilayer degradation by phospholipase A(2). *Biophysical Journal* 1998, 74 (5), 2398–2404.

[662] Milhiet, P.-E.; Gubellini, F.; Berquand, A.; Dosset, P.; Rigaud, J. L.; Le Grimellec, C.; Levy, D., High-resolution AFM of membrane proteins directly incorporated at high density in planar lipid bilayer. *Biophysical Journal* 2006, 91 (9), 3268–75.

[663] Domke, J.; Parak, W. J.; George, M.; Gaub, H. E.; Radmacher, M., Mapping the mechanical pulse of single cardiomyocytes with the atomic force microscope. *European Biophysics Journal* 1999, 28 (3), 179–86.

[664] Friedrichs, J.; Taubenberger, A.; Franz, C. M.; Müller, D. J., Cellular remodelling of individual collagen fibrils visualized by time-lapse AFM. *Journal of Molecular Biology* 2007, 372 (3), 594–607.

[665] Weyn, B.; Kalle, W.; Kumar-Singh, S.; Marck, E. V.; Tanke, H.; Jacob, W., Atomic force microscopy: influence of air drying and fixation on the morphology and viscoelasticity of cultured cells. *Journal of Microscopy* 1998, 189 (2), 172–80.

[666] You, H. X.; Lau, J. M.; Zhang, S. W.; Yu, L., Atomic force microscopy imaging of living cells: a preliminary study of the disruptive effect of the cantilever tip on cell morphology, *Ultramicroscopy* 2000, 82 (1–4), 297–305.

[667] Yokokawa, M.; Takeyasu, K.; Yoshimura, S. H., Mechanical properties of plasma membrane and nuclear envelope measured by scanning probe microscope. *Journal of Microscopy* 2008, 232 (1), 82–90.

[668] Rotsch, C.; Radmacher, M., Drug-induced changes of cytoskeletal structure and mechanics in fibroblasts: an atomic force microscopy study. *Biophysical Journal* 2000, 78 (1), 520–35.

[669] Kim, S. J.; Kim, S.; Shin, H.; Uhm, C. S., Intercellular interaction observed by atomic force microscopy. *Ultramicroscopy* 2008, 108 (10), 1148–51.

[670] Braet, F.; Soon, L.; Kelly, T. F.; Larson, D. J.; Ringer, S. P., Live cell imaging. In *Nanosystem Characterization Tools in the Life Sciences*, Kumar, C. S. S. R., Ed. Wiley-VCH: 2006; pp 309–312.

[671] Puech, P.-H.; Poole, K.; Knebel, D.; Müller, D. J., A new technical approach to quantify cell-cell adhesion forces by AFM. *Ultramicroscopy* 2006, 106 (8–9), 637–44.

[672] Lehenkari, P. P.; Charras, G. T.; Nykänen, A.; Horton, M. A., Adapting atomic force microscopy for cell biology. *Ultramicroscopy* 2000, 82 (1–4), 289–95.

[673] Folprecht, G.; Schneider, S.; Oberleithner, H., Aldosterone activates the nuclear pore transporter in cultured kidney cells imaged with atomic force microscopy. *Pflügers Archiv European Journal of Physiology* 1996, 432 (5), 831–38.

[674] Pandey, V.; Vijayakumar, M. V.; Kaul-Ghanekar, R.; Mamgain, H.; Paknikar, K.; Bhat, M. K., Atomic force microscopy, biochemical analysis of 3T3-L1 cells differentiated in the absence and presence of insulin. *Biochimica et Biophysica Acta* 2009, 1790 (1), 57–64.

[675] Schneider, S.; Sritharan, K.; Geibel, J. P.; Oberleithner, H.; Jena, B., Surface dynamics in living acinar cells imaged by atomic force microscopy: identification of plasma membrane structures involved in exocytosis. *Proceedings of the National Academy of Sciences of the United States of America* 1997, 94 (1), 316–21.

[676] You, H. X.; Yu, L., Atomic force microscopy imaging of living cells: progress, problems and prospects. *Methods in Cell Science* 1999, 21 (1), 1–17.

[677] Li, Q. S.; Lee, G. Y. H.; Ong, C. N.; Lim, C. T., AFM indentation study of breast cancer cells. *Biochemical and Biophysical Research Communications* 2008, 374 (4), 609–13.

[678] Domke, J.; Dannohl, S.; Parak, W. J.; Muller, O.; Aicher, W. K.; Radmacher, M., Substrate dependent differences in morphology and elasticity of living osteoblasts investigated by atomic force microscopy. *Colloids and Surfaces B: Biointerfaces* 2000, 19 (4), 367–79.

[679] Haga, H.; Nagayama, M.; Kawabata, K., Imaging mechanical properties of living cells by scanning probe microscopy. *Current Nanoscience* 2007, 3 (1), 97–103.

[680] Cross, S. E.; Jin, Y. S.; Tondre, J.; Wong, R.; Rao, J.; Gimzewski, J. K., AFM-based analysis of human metastatic cancer cells. *Nanotechnology* 2008, 19 (38), 384003.

[681] Cross, S. E.; Jin, Y. S.; Rao, J.; Gimzewski, J. K., Nanomechanical analysis of cells from cancer patients. *Nature Nanotechnology* 2007, 2 (12), 7803.

[682] Lekka, M.; Laidler, P.; Gil, D.; Lekki, J.; Stachura, Z.; Hrynkiewicz, A. Z., Elasticity of normal and cancerous human bladder cells studied by scanning force microscopy. *European Biophysics Journal* 1999, 28 (4), 312–16.

[683] Gutsmann, T.; Fantner, G. E.; Kindt, J. H.; Venturoni, M.; Danielsen, S.; Hansma, P. K., Force spectroscopy of collagen fibers to investigate their mechanical properties and structural organization. *Biophysical Journal* 2004, 86 (5), 3186–93.

[684] Lee, C. K.; Wang, Y. M.; Huang, L. S.; Lin, S. M., Atomic force microscopy: determination of unbinding force, off rate and energy barrier for protein-ligand interaction. *Micron* 2007, 38 (5), 446–61.

[685] Chtcheglova, L. A.; Haeberli, A.; Dietler, G., Force spectroscopy of the fibrin(ogen) – Fibrinogen interaction. *Biopolymers* 2008, 89 (4), 292–301.

[686] Barattin, R.; Voyer, N., Chemical modifications of AFM tips for the study of molecular recognition events. *Chemical Communications* 2008, 13, 1513–32.

[687] Ebner, A.; Wildling, L.; Kamruzzahan, A. S. M.; Rankl, C.; Wruss, J.; Hahn, C. D.; Holzl, M.; Zhu, R.; Kienberger, F.; Blaas, D.; Hinterdorfer, P.; Gruber, H. J., A new, simple method for linking of antibodies to atomic force microscopy tips. *Bioconjugate Chemistry* 2007, 18 (4), 1176–84.

[688] Lee, I.; Marchant, R. E., Molecular interaction studies of hemostasis: fibrinogen ligand-human platelet receptor interactions. *Ultramicroscopy* 2003, 97 (1–4), 341–52.

[689] Schwesinger, F.; Ros, R.; Strunz, T.; Anselmetti, D.; Guntherodt, H. J.; Honegger, A.; Jermutus, L.; Tiefenauer, L.; Pluckthun, A., Unbinding forces of single antibody-antigen complexes correlate with their thermal dissociation rates. *Proceedings of the National Academy of Sciences of the United States of America* 2000, 97 (18), 9972–77.

[690] Merkel, R.; Nassoy, P.; Leung, A.; Ritchie, K.; Evans, E., Energy landscapes of receptor-ligand bonds explored with dynamic force spectroscopy. *Nature* 1999, 397 (6714), 50–3.

[691] Moy, V. T.; Florin, E. L.; Gaub, H. E., Intermolecular forces and energies between ligands and receptors. *Science* 1994, 266 (5183), 257–59.

[692] Allen, S.; Davies, J.; Dawkes, A. C.; Davies, M. C.; Edwards, J. C.; Parker, M. C.; Roberts, C. J.; Sefton, J.; Tendler, S. J. B.; Williams, P. M., *In situ* observation of streptavidin-biotin binding on an immunoassay well surface using an atomic force microscope. *FEBS Letters* 1996, 390 (2), 161–64.

[693] Ebner, A.; Madl, J.; Kienberger, F.; Chtcheglova, L. A.; Puntheeranurak, T.; Zhu, R.; Tang, J. L.; Gruber, H. J.; Schutz, G. J.; Hinterdorfer, P., Single molecule force microscopy on cells and biological membranes. *Current Nanoscience* 2007, 3 (1), 49–56.

[694] Kienberger, F.; Kada, G.; Mueller, H.; Hinterdorfer, P., Single molecule studies of antibody-antigen interaction strength *versus* intra-molecular antigen stability. *Journal of Molecular Biology* 2005, 347 (3), 597–606.

[695] Allen, S.; Chen, X. Y.; Davies, J.; Davies, M. C.; Dawkes, A. C.; Edwards, J. C.; Roberts, C. J.; Sefton, J.; Tendler, S. J. B.; Williams, P. M., Detection of antigen-antibody binding events with the atomic force microscope. *Biochemistry* 1997, 36 (24), 7457–63.

[696] Soman, P.; Rice, Z.; Siedlecki, C. A., Measuring the time-dependent functional activity of adsorbed fibrinogen by atomic force microscopy. *Langmuir* 2008, 24 (16), 88016.

[697] Rankl, C.; Kienberger, F.; Wildling, L.; Wruss, J.; Gruber, H. J.; Blaas, D.; Hinterdorfer, P., Multiple receptors involved in human rhinovirus attachment to live cells. *Proceedings of the National Academy of Sciences of the United States of America* 2008, 105 (46), 17778–83.

[698] Zhang, X.; Chen, A.; De Leon, D.; Li, H.; Noiri, E.; Moy, V. T.; Goligorsky, M. S., Atomic force microscopy measurement of leukocyte-endothelial interaction. *American Journal of Physiology – Heart and Circulatory Physiology* 2004, 286 (1), 359–67.

[699] Lamontagne, C.-A.; Cuerrier, C.; Grandbois, M., AFM as a tool to probe and manipulate cellular processes. *Pflügers Archiv European Journal of Physiology* 2008, 456 (1), 61–70.

[700] Heinz, W. F.; Hoh, J. H., Spatially resolved force spectroscopy of biological surfaces using the atomic force microscope. *Trends In Biotechnology* 1999, 17 (4), 143–50.

[701] Yersin, A.; Steiner, P., Receptor trafficking and AFM. *Pflügers Archiv European Journal of Physiology* 2008, 456 (1), 189–98.

[702] Kim, H.; Arakawa, H.; Hatae, N.; Sugimoto, Y.; Matsumoto, O.; Osada, T.; Ichikawa, A.; Ikai, A., Quantification of the number of EP3 receptors on a living CHO cell surface by the AFM. *Ultramicroscopy* 2006, 106 (8–9), 652–62.

[703] Gad, M.; Itoh, A.; Ikai, A., Mapping cell wall polysaccharides of living microbial cells using atomic force microscopy. *Cell Biology International* 1997, 21 (11), 697–706.

[704] Verbelen, C.; Christiaens, N.; Alsteens, D.; Dupres, V.; Baulard, A. R.; Dufrêne, Y. F., Molecular mapping of lipoarabinomannans on mycobacteria. *Langmuir* 2009, 25 (8), 4324–27.

[705] Gilbert, Y.; Deghorain, M.; Wang, L.; Xu, B.; Pollheimer, P. D.; Gruber, H. J.; Errington, J.; Hallet, B.; Haulot, X.; Verbelen, C.; Hols, P.; Dufrêne, Y. F., Single-molecule force spectroscopy and imaging of the vancomycin/D-Ala-D-Ala interaction. *Nano Letters* 2007, 7 (3), 796–801.

[706] Dufrêne, Y.; Hinterdorfer, P., Recent progress in AFM molecular recognition studies. *Pflügers Archiv European Journal of Physiology* 2008, 456 (1), 237–45.

[707] Chtcheglova, L. A.; Waschke, J.; Wildling, L.; Drenckhahn, D.; Hinterdorfer, P., Nano-scale dynamic recognition imaging on vascular endothelial cells. *Biophysical Journal* 2007, 93 (2), L11–3.

[708] Stroh, C. M.; Ebner, A.; Geretschläger, M.; Freudenthaler, G.; Kienberger, F.; Kamruzzahan, A. S. M.; Smith-Gill, S. J.; Gruber, H. J.; Hinterdorfer, P., Simultaneous topography and recognition imaging using force microscopy. *Biophysical Journal* 2004, 87 (3), 1981–90.

[709] Kienberger, F.; Ebner, A.; Gruber, H. J.; Hinterdorfer, P., Molecular recognition imaging and force spectroscopy of single biomolecules. *Accounts of Chemical Research* 2006, 39 (1), 29–36.

[710] Grandbois, M.; Dettmann, W.; Benoit, M.; Gaub, H. E., Affinity imaging of red blood cells using an atomic force microscope. *Journal of Histochemistry & Cytochemistry* 2000, 48 (5), 719–24.

[711] Lee, G. U.; Chrisey, L. A.; Colton, R. J., Direct measurement of the forces between complementary strands of DNA. *Science* 1994, 266 (5186), 771–73.

[712] Bonin, M.; Zhu, R.; Klaue, Y.; Oberstrass, J.; Oesterschulze, E.; Nellen, W., Analysis of RNA flexibility by scanning force spectroscopy. *Nucleic Acids Research* 2002, 30 (16), e81.

[713] Mitsui, K.; Hara, M.; Ikai, A., Mechanical unfolding of alpha(2)-macroglobulin molecules with atomic force microscope. *FEBS Letters* 1996, 385 (1–2), 29–33.

[714] Oberhauser, A. F.; Marszalek, P. E.; Erickson, H. P.; Fernandez, J. M., The molecular elasticity of the extracellular matrix protein tenascin. *Nature* 1998, 393 (6681), 181–85.

[715] Rief, M.; Gautel, M.; Oesterhelt, F.; Fernandez, J. M.; Gaub, H. E., Reversible unfolding of individual titin immunoglobulin domains by AFM. *Science* 1997, 276 (5315), 1109–12.

[716] Brockwell, D. J., Force denaturation of proteins – an unfolding story. *Current Nanoscience* 2007, 3 (1), 3–15.

[717] Borgia, A.; Williams, P. M.; Clarke, J., Single-molecule studies of protein folding. *Annual Review of Biochemistry* 2008, 77, 101–25.

[718] Carrion-Vazquez, M.; Oberhauser, A. F.; Fowler, S. B.; Marszalek, P. E.; Broedel, S. E.; Clarke, J.; Fernandez, J. M., Mechanical and chemical unfolding of a single protein: a comparison. *Proceedings of the National Academy of Sciences of the United States of America* 1999, 96 (7), 3694–99.

[719] Ng, S. P.; Rounsevell, R. W. S.; Steward, A.; Geierhaas, C. D.; Williams, P. M.; Paci, E.; Clarke, J., Mechanical unfolding of TNfn3: the unfolding pathway of a fnIII domain probed by protein engineering, AFM and MD simulation. *Journal of Molecular Biology* 2005, 350 (4), 776–89.

[720] Oberhauser, A. F.; Hansma, P. K.; Carrion-Vazquez, M.; Fernandez, J. M., Stepwise unfolding of titin under force-clamp atomic force microscopy. *Proceedings of the National Academy of Sciences of the United States of America* 2001, 98 (2), 468–72.

[721] Schlierf, M.; Li, H. B.; Fernandez, J. M., The unfolding kinetics of ubiquitin captured with single-molecule force-clamp techniques. *Proceedings of the National Academy of Sciences of the United States of America* 2004, 101 (19), 7299–7304.

[722] Bippes, C. A.; Janovjak, H.; Kedrov, A.; Müller, D. J., Digital force-feedback for protein unfolding experiments using atomic force microscopy. *Nanotechnology* 2007, 18 (4), 044022.

[723] Schäffer, T. E., Calculation of thermal noise in an atomic force microscope with a finite optical spot size. *Nanotechnology* 2005, 16 (6), 664–70.

[724] Chinga-Carrasco, G.; Kauko, H.; Myllys, M.; Timonen, J.; Wang, B.; Zhou, M.; Fossum, J. O., New advances in the 3D characterization of mineral coating layers on paper. *Journal of Microscopy* 2008, 232 (2), 212–24.

[725] Di Risio, S.; Yan, N., Characterizing coating layer z-directional binder distribution in paper using atomic force microscopy. *Colloids and Surfaces A: Physicochemical and Engineering Aspects* 2006, 289 (1–3), 65–74.

[726] Sadaie, M.; Nishikawa, N.; Ohnishi, S.; Tamada, K.; Yase, K.; Hara, M., Studies of human hair by friction force microscopy with the hair-model-probe. *Colloids and Surfaces B: Biointerfaces* 2006, 51 (2), 120–29.

[727] Swift, J. A.; Smith, J. R., Atomic force microscopy of human hair. *Scanning* 2000, 22 (5), 310–18.

[728] Breakspear, S.; Smith, J. R.; Luengo, G., Effect of the covalently linked fatty acid 18-MEA on the nanotribology of hair's outermost surface. *Journal of Structural Biology* 2005, 149 (3), 235–42.

[729] Chen, N.; Bhushan, B., Morphological, nanomechanical and cellular structural characterization of human hair and conditioner distribution using torsional resonance mode with an atomic force microscope. *Journal of Microscopy* 2005, 220 (2), 96–112.

[730] Knight, S.; Dixson, R.; Jones, R. L.; Lin, E. K.; Orji, N. G.; Silver, R.; Villarrubia, J. S.; Vladar, A. E.; Wu, W. L., Advanced metrology needs for nanoelectronics lithography. *Comptes Rendus Physique* 2006, 7 (8), 931–41.

[731] Foucher, J.; Ernst, T.; Pargon, E.; Martin, M. Critical dimension metrology: perspectives and future trends. http://dx.doi.org/10.1117/2.1200811.1345.

[732] Kwon, J.; Kim, Y.-S.; Yoon, K.; Lee, S.-M.; Park, S.-i., Advanced nanoscale metrology of pole-tip recession with AFM. *Ultramicroscopy* 2005, 105 (1-4), 51–56.

[733] Gupta, B. K.; Young, K.; Chilamakuri, S.; Menon, A. K., Head design considerations for lower thermal pole tip recession and alumina overcoat protrusion. *Tribology International* 2000, 33 (5-6), 309–14.

[734] Choi, M.; Yang, J. M.; Lim, J.; Lee, N.; Kang, S., Measurement and analysis of magnetic domain properties of high-density patterned media by magnetic force microscopy, IEEE Transactions on Magnetics 2009, 45 (5), 2308–11.

[735] Breakspear, S.; Smith, J. R.; Nevell, T. G.; Tsibouklis, J., Friction coefficient mapping using the atomic force microscope. *Surface and Interface Analysis* 2004, 36 (9), 1330–34.

[736] Nie, H.-Y.; McIntyre, N. S., A simple and effective method of evaluating atomic force microscopy tip performance. *Langmuir* 2001, 17 (2), 432–36.

[737] Bolhuis, T.; Pasop, J. R.; Abelmann, L.; Lodder, J. C. *Scanning probe microscopy markup language*, Scanning Tunneling Microscopy/Spectroscopy and Related Techniques: 12th International Conference. AIP: Eindhoven, 2003; pp 271–8.

[738] Horcas, I.; Fernandez, R.; Gómez-Rodriguez, J. M.; Colchero, J.; Gomez-Herrero, J.; Baró, A. M., WSXM: A software for scanning probe microscopy and a tool for nanotechnology. *Review of Scientific Instruments* 2007, 78 (1), 013705.

Index

Italic numbers at end of list indicate reference to figures.